DATE DUE

Estimating Electrical Construction

By
Edward J. Tyler

Craftsman Book Company
6058 Corte del Cedro, Box 6500, Carlsbad, CA 92008

Acknowledgements

The author wishes to express his appreciation to the following companies and organizations for furnishing materials used in the preparation of various portions of this book.

American Arbitration Association, 140 West 51st Street, New York, New York 10020
Appleton Electric Company, 1701-59 Wellington Avenue, Chicago, Illinois 60657
Johns-Manville Sales Corporation, Ken-Caryl Ranch, Denver, Colorado 80217
Leviton Manufacturing Company, Inc., 59-25 Little Neck Parkway, Little Neck, New York 11362
Lithonia Lighting, Box A, Conyers, Georgia 30207
Marvin Electric Manufacturing Company, 6100 S. Wilmington Avenue, Los Angeles, Calif. 90001
Square D Company, Executive Plaza, Palatine, Illinois 60067
Trade Service Publications, Inc., 10996 Torryana Road, San Diego, California 92121
Westinghouse Electric Corporation, Lighting Division, P. O. Box 824, Vicksburg, Miss. 39180

This book is dedicated to all active and student estimators.

Library of Congress Cataloging in Publication Data
Tyler, Edward J.
    Estimating electrical construction.

    Includes index.
    1. Electric engineering--Estimates.   I. Title.
TK435.T93 1983 621.319'24   82-23663
ISBN 0-910460-99-X

©1983 Craftsman Book Company

Edited by Sam Adrezin

Photographs by Mark C. Tyler

Third printing 1985

Contents

Chapter 1

Electrical Contracting

Electrical contracting is a relatively new trade. But it has become a very detailed and exacting discipline where most contractors specialize in certain types of projects or services. Few firms handle all types of electrical work, though many can perform more than one specialty. But most concentrate on developing their skills and building their reputation in one area.

In the early days of electrical contracting, installation practices were poor. Still, the demand for electricity increased rapidly. The young trade associations joined with other industry groups to develop product and installation standards. In the early 1900's their efforts produced positive results when the National Fire Protection Association developed the National Electrical Code.

The National Electrical Code aims to protect the public. It is revised about every three years. The revisions are the result of code committee studies, better products, advanced designs, and improved installation procedures. Keeping up with all the latest developments makes the role of the electrical estimator much more difficult. But it also makes his work better, safer, and more enjoyable.

The Electrical Estimator

Many people in the electrical construction industry quote prices. These include company owners, managers, engineers, supervisors, electricians and salespeople. Some quote vague "ball park" figures. But most use charts, tables, measurements and calculations to arrive at an accurate estimate. They've found that accuracy is the key to success, even for the simplest job.

The following were among the first to estimate electrical jobs:

Electricians who had field experience, were very good with details and were experts at interpreting the newly developed National Electrical Code.

Electrical engineers who were needed to design the systems.

Salesmen who had the product knowledge and some idea of installation techniques.

Specialists from government electrical jobs. Their experience came from procurement, quality control, and on-site inspection.

Technicians from the telephone companies, alarm system manufacturers, and electrical equipment manufacturers.

As the industry grew more complex, the need for specialists increased. Estimators were needed who could accurately determine job costs for various types of projects. The first electrical estimators were the master electricians with many years of field experience. They could read drawings and understand contract specifications. These abilities are still essential in preparing an accurate electrical estimate.

The master electricians knew how to design electrical layouts. They would draw the electrical systems on the floor plans, showing the locations of outlets, switches and lighting fixtures. These drawings were then used as installation guides—nothing more. The price of the job was an educated guess by either the electrical contractor or the layout man.

Eventually the estimators learned to do material take-offs from their electrical drawings. These drawings then became a companion to the architectural drawings.

Today, electrical estimators use intricate estimating procedures. But most of these practices can be easily taught. Contractors no longer depend on the master electrician for predicting costs. The trend now is to train new estimators. These include junior electrical engineers specializing in construction estimating, and high school and college graduates with a background in mathematics and mechanics. With a few years of on-the-job training, they can become competent and successful electrical construction estimators.

Rewards

Electrical estimating offers many rewards. One is good pay. Most companies offer a fair beginning salary with periodic increases. Some base the salary on a percentage of the rate paid to the craftsmen who do the actual field installations. The percentage is low in the beginning, but increases with experience and competence. The estimator works on a salary, either weekly or bimonthly, including fringe benefits.

Advancement opportunities are excellent for the progressive estimator. As the planner of new work, the estimator makes important contributions to the success of the company. Many estimators eventually become contractors and owners of their own companies.

Working conditions are good. But time is a big problem. The estimator is always battling the clock. All bid dates are set by the owner.

Carefully go over all of the drawings for details that will affect the electrical installation. Many times other drawings will indicate the required location for the electrical item.
Figure 1-1

Some can be changed to fit the customer's schedule or to allow for design changes. This can be frustrating, but with good planning and hard work these problems can be overcome. The take-off must be finished on time. That might mean spending a few additional hours at the office. But it will be worth it when you see the completed project.

The Design Team

Generally the design team consists of an architect, a structural engineer, a mechanical engineer and an electrical engineer. At times, other specialists are needed. The leader of the design team is the architect.

The architect: The architect designs and supervises the project. He develops the design according to the owner's requirements. In most cases the owner selects the architect and places him in full charge of the job. The architect then enlists the aid of consulting engineers for advice on specific technical phases of the design. Schematics are prepared for the project, and from these the consulting engineers can start their work.

The architect prepares the construction documents. These include the advertisement for bids, the instructions to the bidders, the contract, the general and special conditions to the contract, the project specifications, the project drawings and the guidelines for administration of the contract.

The mechanical engineer: Mechanical engineers (M.E.) design systems to generate, transport, and convey heat and power. The M.E. coordinates his designs with the structural and architectural designs. In buildings, the M.E. designs the heating, plumbing, drainage, ventilation, exhaust and air conditioning systems.

The electrical engineer: Electrical engineers (E.E.) design systems to convey and generate electricity and artificial lighting and provide communications networks. The E.E. must work closely with the architect, the structural engineer and the mechanical engineer.

Additional engineering duties: The design team may be responsible for on-site inspections as the project progresses. Many contracts between the architect and the consulting engineers require inspections to ensure that installation is made in accordance with specifications and drawings.

The consultants check the shop drawings prepared by the manufacturer or by the subcontractors. They must also check material lists and certain catalog data submitted for approval by the contractor. The consultant will check these against the design criteria to be sure the specifications are met.

Builder-Designer
Often the builder is also the designer. Most states permit a builder to do the design work for a project that he will construct, though restrictions may prohibit him from designing work for others.

Frequently the builder is part of the design team. He can contribute valuable information about materials and installation methods and may suggest alternatives.

The builder may be the developer and will organize the design team. He will usually take an active part in the overall design, but will still put the architect in charge. In some cases the builder may administer the contract and set the construction schedule.

Sometimes the owner might contract with the builder to take full charge of the project. Most owners don't have the time or the background to deal with construction problems. It's usually easier, faster, and cheaper to assign the project to a dependable, well-known builder.

Chapter 2

Estimating Procedures

This chapter outlines the estimating process in general terms and identifies some general principles that apply to every electrical estimate. It serves as an introduction to the later chapters which look in detail at the main points that are sketched with a broad brush here.

Most estimates are in two parts: material and labor. These are the foundation of the cost estimate and must be accurate. Most of the direct and indirect costs will be calculated from the material and labor figures.

Material

A careful quantity take-off is essential. Current wholesale prices, quotations for special items such as lighting fixtures and distribution equipment, and subcontract prices for special systems will determine the material costs of the project. Materials usually account for about sixty percent of the direct job costs.

Materials are the most predictable part of the estimate. Most expensive items should be covered by firm quotations. Rough-in items are purchased during the early phases of construction, and trim and finish are purchased last. The largest money risks are thus avoided.

Labor

Labor has always been hard to predict because there are so many factors that can affect production. These include poor job site supervision, badly scheduled material deliveries, incorrect materials, inadequate tooling, lack of proper equipment, bad weather, strikes by on-site trades or manufacturing, poor working relations with other trades at the site, adverse site conditions, unrealistic inspectors, and many unforeseen circumstances. Experience is the most helpful tool in understanding the conditions that might affect labor costs.

Organize Job Costs

Early estimators studied their previous jobs to find factors that could be applied to new bids. The use of company records helped establish material and labor codes. Material could be classified into installation categories and coded for job cost accounting and purchase control. Labor was classified by each installation procedure and then coded. After using these codes for several jobs, the estimator had a set of standards that could be applied to new work. Job costs could be predicted more accurately. The material and labor codes formed a firm foundation for project estimating.

Adjustments were made as experience with the new system was gained. Accuracy increased. Today some estimators still use company records for their estimates.

Most estimators are reluctant to reveal their estimating procedures. They want to shield their valuable information from competitors. That's understandable. They've developed methods which benefit their company, and they don't want others capitalizing on their secrets.

Still, even with the best possible procedures for predicting job costs, many good estimators lose jobs to bidders who don't adhere to trade ethics and standards.

In most areas of the country hundreds of contracting firms compete for the same type of work. Many jobs are bid by dozens of contractors. Some use the "shotgun" approach to get the bid price. That is, they take a wild guess at what the price might be, pressure the general contractor into revealing the lowest bids he's received, then undercut those bids by a slim margin. In many cases, the general contractor freely offers the bidding prices he receives to selected subcontractors.

Some don't bid the job in accord with the project specifications. Either they take exception to certain requirements or they gamble that they can get away with altering the wiring systems to reduce the overall installation costs. Most of these estimators don't survive. They usually outsmart themselves in the long run. But they cause problems and waste time for the other estimators.

Labor Units

After World War I, specialty contractor trade associations became a strong influence in the construction industry. They were established to promote the best possible business climate. These associations helped promote professionalism and better bidding practices in the industry.

Company owners were the prime movers of the associations. Member companies furnished information about their estimating procedures. This was information on general practices—not the estimator's secrets. After analyzing the practices submitted, the association members adopted a standard format for estimating electrical work. They decided that a schedule of installation values for manhours of labor was needed. Members were asked to evaluate certain materials and determine installation times. Eventually enough information was collected for the publication of a manual that could be used by all members of the association.

The manual contains *labor units* that can be applied to a *take-off* or a bill of material prepared from the proposed contract documents. This manual has been recognized as the electrical industry standard for installation labor since the early 1920's. Most government and construction industry segments recognize the manual as the standard also.

The manual is revised periodically to keep up with new products and new installation techniques. It is the property of the association and is loaned to member companies for their exclusive use. Most firms using the manual adjust some of the labor units for local conditions and for their particular needs.

For information concerning the Manual of Labor Units, contact the NECA chapter in your area. They will be listed in the telephone directory as National Electrical Contractors Association. You may also contact their national headquarters by writing to:

National Electrical Contractors Association, Inc.
7315 Wisconsin Avenue
Washington D.C. 20014
(301) 657-3110

Unit Costs

Many successful contracting firms have developed estimating techniques other than those advocated in the manual to fit their special needs. When a firm specializes in an area of work where a take-off might list only a few items, it might be to their advantage to establish a simple unit cost schedule for pricing jobs. In fact, some bids require unit prices for all work under contract.

Highway projects are an example. Total cost for these projects can be high, perhaps several million dollars. The bid form might list from a few to dozens of items that must be priced and extended separately. The bid might also call for certain *lump sum* prices. Lump sum items are priced by making a detailed take-off of all of the material and labor specified in that lump sum bid item. The bid form usually lists the quantity for each unit. This way the unit cost can be extended to a total cost for that item. Then all of the extended bid items are added up for the total project cost. The owner's representative may have the authority to adjust the quantities of the bid item, either up or down, when the project is underway. *The unit cost must be as accurate as possible.*

Another typical use of unit costs is in commercial developments such as shopping centers and office buildings. The contract bid form may call for a lump sum price for the building shell, the building services, and provisions for future tenant improvements. The interior is usually designed specifically for the tenant's needs as the leases are made. When the project is up for bid, most bid forms call for unit costs that can be applied to the cost of completing the interiors. In many cases the tenant has the right to arrange for the interior work to be done by someone other than the project contractor. The project contractor needs accurate prices for the original building shell. This allows an accurate estimate for all or just part of the project.

Electronic Estimating

Automated estimating systems are available from companies that sell estimating aids to contractors. Special probes are used to scale footage and count items during the take-off process. These aids or tools can be found in small, inexpensive packages or in large units with magnetic discs for programming estimate pricing.

Some automated estimating systems can handle a range of estimating programs and store thousands of coded items. Many systems can be easily adapted to payroll accounting and inventory control. Some store data on magnetic discs and are programmed for the desired operation. After all material and labor items have been taken off, the items are coded and processed into the system. The operation is usually very quick and simple. The printout reports can be simple or highly detailed and complex.

Computerized estimating systems are used all over the country. At one time the computer hardware was too expensive for the average contractor. The computer language was complicated and a programmer was needed to set up the software. Now there are many inexpensive computers available that require little special training to use.

Some computer estimating companies have set up subscription services for contractors. They've established central locations where the computer and programs are maintained with up-to-date pricing for material and labor. Computer terminals are placed in the subscribing contractor's office, and the estimator is shown how to operate the system. When the take-off has been processed into the computer, the estimator signals the central computer to receive the specially coded data. Almost instantly, the computer provides the estimator with a detailed printout of pricing data. The program is as accurate as the estimator's take-off. The calculations from the computer are mathematically correct. Many safeties are built into the program to reduce the possibility of errors.

The breakdowns are so complete on the printouts that materials can be readily purchased, equipment requirements can be scheduled, and the crew size can be established for the installations.

But materials such as lighting fixtures, distribution equipment, communications systems and other specially designed items are usually not priced by the computer. These must be priced by the manufacturer's representatives or by the wholesale suppliers who research the project requirements and prepare quotations on those items.

Other costs must be added after the computer has priced the material. These include the *direct* and *indirect* overhead costs. The special and general conditions in the specs influence these costs. There may be additional requirements for equipment hookup of items furnished under other sections of the contract but performed by the electrical contractor. Most of these requirements are detailed in the contract documents. They include permits, additional insurance, travel pay, utility fees or any other costs assessed to the job.

A computer estimate must be adjusted to compensate for local purchasing levels, weather and labor productivity. The contractor must determine the amount of adjustment needed to ensure the right price and the desired profit level.

Computers provide a speedy, mathematically correct way to price a job. But the accuracy of the take-off—the right quantities and types of material and the difficulty of the installation—is still the responsibility of the estimator. It's almost impossible for two or more estimators to arrive at the same material quantities. But computer costing can make two or more estimates for the same project very close. The difference will be in the supplier's quotations, the factoring for local conditions, the calculation of special contract provisions and the amount of overhead and profit. No two contractors will evaluate these items in the same way or come up with the same figures.

Computers are a big aid for many electrical

estimators. But they don't replace judgement and haven't put any electrical estimators out of business. Don't get the idea that every electrical estimator needs a computer or that with a computer you don't really need to understand estimating to produce an accurate bid. Most jobs, especially smaller jobs, don't require a computer. And it's still possible to make big mistakes even with the help of the most sophisticated computer and estimating program. Computers are an estimating aid like an adding machine or architect's scale. But no machine, no matter how reliable and complex, will ever dictate what electrical contractors charge for their services.

Include All Costs
Every accurate estimate includes all costs necessary to complete the contract in accordance with the specifications. A bad bid, whether too low or too high, will always be a loser. If it's too low, it may cost the contractor all or most of his profit. He may even suffer a loss on the job. If the bid is too high, the contractor won't get the job. He'll have wasted time and effort that could have been used more productively elsewhere.

Every bid must be a good bid, even though it's not always the lowest bid. Sometimes the contractor who has made a mistake and bid the job too low can get out of the job. When this happens the next higher bidder may be offered the work.

Winning one out of every eight bids is a good average. A contractor who gets most of the jobs he bids may be in for trouble. Chances are he's doing something wrong and will pay for it eventually. When four to ten bidders normally bid the average job, don't try to be low every time.

Remember, a good bid includes all of the costs that can be reasonably identified in the bidding documents, proper identification of the materials and installation requirements, and correct extensions and additions throughout the total take-off.

The Plan Take-off
The *plan take-off* lists all material and work items indicated on the contract drawings. Check the electrical drawings for all lighting fixtures and the provisions needed to install those fixtures. Your take-off should include the following items:
• Lighting fixture lamps (light bulbs)

• Wiring devices (switches, receptacles, cover plates, special outlets, time switches, photo cells).
• Distribution equipment (metering lighting panels, main electrical panel, transformers, motor control centers, bus duct, terminal cabinets).
• Conduit (rigid, EMT flexible, PVC).
• Conduit fittings (elbows, locknuts, insulated bushings, insulated ground bushings, straps, pipe clamps, hangers, condulets, expansion fittings, nipples).
• Wire (various sizes, types of insulation, splicing materials).
• Trenching and backfill.
• Excavation for manholes, pull boxes and light pole foundations.
• Encasement material.
• Equipment connections (A/C units, boilers, air handling systems, exhaust systems, mechanical equipment control centers, pumping, conveying).

Your estimate must include the cost of every item to be purchased, handled, installed, tested and inspected in accordance with the contract requirements.

After completing the plan take-off and adding up the material quantities, transfer the take-off from the work sheets to the pricing sheets. Make the entries on the pricing sheets clear and legible. Try to establish a definite routine that you use on every estimate. That makes omitting something less likely. For example, place lighting fixtures and conduit in a certain order on the pricing sheets and in the pricing sheet index. Do it the same way on each estimate. A good, well-planned format allows you to review an estimate quickly and easily without running back to the plans or the pricing sheet for clarification.

Indexing The Estimate
Set up an indexing system that can be used from one estimate to the next. The key is to keep the process as simple as possible. The following is a typical index.
1. Switchgear:
Metering equipment, distribution switchboards, motor control centers, lighting and power panels, transformers, terminal cabinets, feeder circuit breakers.
2. Lighting Fixtures:
Plaster frames, remote ballasts, stem hangers, spacers, couplings, tandem units, suspension systems, floodlight poles, brackets, special

backing, dimming equipment, interference suppressors, diodes, in-line fusing, chain hanger material.

3. Lamps:

Incandescent, fluorescent, mercury vapor, high pressure sodium, low pressure sodium.

4. Conduit and Fittings:

Rigid, flexible, EMT, PVC, connectors, couplings, elbows, locknuts, bushings, straps, hangers, conduit bodies.

5. Boxes and Box Fittings:

Fixture outlet boxes, single gang boxes for lighting switches, receptacles, special outlets, multi-ganged plates for double or more wiring devices, boxes for junctions.

6. Wire and Cable:

Control wire, building wire, feeder cable, splices, connectors.

7. Wiring Devices:

Lighting switches, receptacles, special outlets, plates, covers.

8. Safety Switches:

Size and type (indoor-outdoor, fused-nonfused, general–heavy duty, horsepower rated, magnetic starters, combination switch-starter, manual motor starters, wireway-gutter, large pull boxes).

9. Equipment Hookup:

Air handling, exhaust, air conditioning, boilers, chillers, air compressors, automatic doors, conveyors, pumps, mechanical equipment control centers and devices.

10. Trenching and Excavation:

Various trench sizes, excavating for manholes and pull boxes, concrete encasements, sand, backfilling, cleanup, saw cutting of pavement, core drilling.

11. Demolition:

Disconnection of services and equipment; removal of fixtures, conduit, wire.

The Bid Summary Sheet

There are many ways to prepare a bid summary sheet. Some contractors design their own bid summary sheets to fit their business needs. Others use a standard summary sheet produced by publishers of business forms. Bid summary forms can range from the very simple to the very detailed. Figure 2-1 is a typical bid summary sheet.

The Spread Sheet

To maintain control of the many quotations received just prior to bid time, estimators use a *spread sheet* to compare prices. (Figure 2-2.)

The left column lists the fixture types. These are found on the fixture schedule or in the specifications. Next to the fixture type is the quantity of fixtures used. The wholesale supplier or distributor is listed at the head of the pricing column. Prices are listed when received. After all of the prices have been listed, they are extended by multiplying the fixture quantity by the price of each fixture.

Lump sums are listed in larger spaces and indicated by a brace. The spread sheet can be extended to include more fixture types by taping on additional sheets. A similar spread sheet can be used for distribution equipment and other special material items. When all of the special materials have been priced and the best prices have been determined, transfer the prices to the estimate end sheet (bid summary).

The Checker

Someone must check the completed estimate to make sure there are no mathematical errors. This *checker* should go over every extension on the pricing sheets and all the figures on the summary sheet. Each extension and addition should be verified with a check mark to show that it was checked.

The checker should never make a correction without the approval of the estimator. The figure should be marked for the estimator to check later. This way, the estimator maintains responsibility for, and control of, the estimate.

Delays

Many estimates are completed days ahead of the bid date. Sometimes the bid date is postponed by the owner's representative. Occasionally this is to allow time to make changes in the bidding documents because of discrepancies found by a contractor or supplier. The estimator should be able to set the take-off aside, start and complete other estimates, then return to the take-off—all with complete control. To do this, he must have confidence in his bids. That confidence comes from following a controlled take-off procedure that allows for periodic improvement.

Estimating Forms

Contractors and estimators have designed many special estimating forms. Some examples and their uses are shown at the end of this chapter. Blank copies are provided at the end of the book.

Bid Summary

Job: RIVER HOUSE CLUB   Estimate No.: M 386
Location: PADDLEWHEEL LANE   Bid Date: 4/1
Division: 16A   Time: 2:00 P.M.
Estimator: ED   Checker: SUE

	Description	Material		Deducts		Labor
1	GEAR	1 200.00 ✓				14.00 ✓
2	FIXTURES	5 678.00 ✓				123.25 ✓
	LAMPS	263.00 ✓				6.00 ✓
3	CONDUIT/FITTINGS	2 875.00 ✓				98.60 ✓
4	BOXES, WIRE	1 999.00 ✓				78.00 ✓
5	DEVICES, SAF. SW	334.00 ✓				10.10 ✓
6	HOOKUP	50.00 ✓				4.00 ✓

	Material			Labor
Sub Total	12 399.00 ✓			333.95
Sales Tax 5 %	620.00 ✓	Supervision % 10		33.40
Sub Total	13 019.00 ✓	Total Hours		367.35
Labor	6 980.00 ✓	Rate		13.20
Tools	242.00 ✓	Sub Total		4849.02 ✓
Miscellaneous	100.00	Fringes 2.25		826.54 ✓
Permits/Fees	200.00	Taxes 20		969.80 ✓
Sub Total	20 541.00 ✓	Sub Total		6645.36 ✓
Subcontracts	—0—	Factor		334.52 ✓
Travel Expense	—0—	Total		6979.88 ✓
Sub Total	20 541.00 ✓			
Deducts		Addendums		1, 2, 3
Net Cost		Alternates		NONE
Overhead		Job Duration		180 CAL. DAYS
Sub Total		Penalty		$35.00/DAY
Profit		Type Const.		NEW-E
Contingency				
Bond				
Selling Price				

Figure 2-1

Spread Sheet

Job: RIVER HOUSE CLUB   **Estimate No.:** M386

Material: LIGHTING FIXTURES   **Date:** 4/1

		GESCO ART		WESCO JUDY		GRAYBAR BILL		STAR MIKE		VALLEY OMAR	
A	12	30.25	363.00	33.00	396.00	28.60	343.20	36.50	438.00	27.00	324.00
B	8	10.60	84.80	10.60	84.80	10.60	84.80	11.00	88.00	10.90	87.20
C	50	75.00	3750.00	74.00	3700.00	75.00	3750.00	76.00	3800.00	10.00	*500.00
D	16	18.00	288.00	17.95	287.20	18.00	288.00	18.00	288.00	19.00	304.00
E	20	8.00	160.00	10.00	200.00	10.00	200.00	9.00	180.00	9.50	190.00
F	2	15.00	30.00	18.00	36.00	17.00	34.00	19.00	38.00	17.00	34.00
G	1	—	140.00	—	137.50	—	142.00	—	155.00	—	148.00
H	4										
J	14		650.00		645.00		648.00		670.00		648.00
K	5										
L	1										
M	0	—	—	—	—		8.90	—	—	8.60	—
N	11	4.00	44.00	6.00	66.00	4.50	49.50	3.80	41.80	4.00	44.00
P	2	12.00	24.00	16.00	32.00	8.00	16.00	14.00	28.00	12.00	24.00
R	6	20.00	120.00	20.00	120.00	20.00	120.00	21.00	126.00	24.00	144.00
S	12	1.00	12.00	1.05	12.60	1.00	12.00	2.00	24.00	1.50	18.00
T	2	6.00	12.00	7.00	14.00	6.00	12.00	6.00	12.00	6.00	12.00
			5677.80		5731.10		5699.50		5888.80		2477.20

* ERROR

Figure 2-2

Large operations with more than one estimator devise many special forms to ensure uniformity. Some forms enable the estimator to determine if the proposed project is right for the contractor. A good form includes a check list with enough categories to help the contractor decide if he should bid a job. Sometimes a project is not as good as it looks. Carefully study each prospective job to determine if it will be good for the company and if the proposed work will coincide with work in progress. The company's work experience, the availability of qualified employees, tools and equipment, and the financial impact are but a few items to consider before assembling an estimate.

Forms are available for most estimating needs. Good forms make the job of estimating easier to control by providing good clear records of each phase. They should be simple and easy to use. Color coding different forms helps provide quick access and easy identification.

Trade associations can provide their members with a variety of estimating forms. Determine which forms you'll require, then contact your local trade association office to get the forms you need.

Scope of Work

Estimate No. _____

Job _RIVER HOUSE CLUB_   Bid Date _4/1_
Location _PADDLEWHEEL LANE_   Estimator _ED_
Bids To _OWNER_   Location _1400 ST. MARY'S ST._   Time _2 P.M._

1. Design Team
Architect — BROWN & ASSOC.
Engineer — AL WHITE
Agency — PRIVATE
Owner — HAM CO.

2. Construction
Building — 1500 Sq.Ft. - B
Walls — CONCR. BLOCK
Ceilings — BEAM
Floors — TILE/CARPET

3. Quotations
Switchgear — 1200A 3∅ 208 V.
Generator — NONE
Alarm Systems — N/C
U.F. Duct — NONE
Communication Systems — N/C
Cable Tray — NONE
Fixtures — VARIETY
Telemetry — NONE

4. Specified Items
Conduit — STANDARD CODE
Wire — THHN
Switches — SPEC GRADE
Receptacles — SPEC GRADE
Dimming Equipment — INDIVIDUAL
Motor Control — BY MECH.
Manholes — NONE
Concrete — BY CONTR.

5. Related Work
Temporary — BY OWNER
Control Wiring — BY MECH.
Starters — BY MECH.
Painting — NONE
Service Cable — UTILITY CO.
Pole Bases — BY CONTR

6. Site Conditions
Excavation — FAIR TO EASY
Access — GOOD
Utilities — AT SITE
Security — NONE
Pave Cutting — NONE
Pave Patch — NONE
Demolition — NONE

Project Selection Checklist

Job _RIVER HOUSE CLUB_ Bid Date _4/1_

Location _PADDLEWHEEL LANE_ Estimator _ED_

Financial

1. Approximate Cost of Work _25,000, 00_
2. Bonding Required _NO_
3. Progress Payments _MONTHLY_
4. Retention _10_ %
5. Delay Penalties _35.00/DAY_
6. _____
7. _____

Project Type

1. Residential _____
2. Commercial _✓_
3. Industrial _____
4. Institutional _____
5. Underground _____
6. Overhead _____
7. Waterfront _____
8. High Voltage _____
9. Communications _____
10. _____
11. _____
12. _____

Bid Documents

1. Complete Plans _✓_
2. Complete Specs. _✓_
3. Reduced Plans _FULL SIZE_
4. Plan Deposit _NONE_
5. Public Bid _NO_
6. Sublisting _NO_
7. Prequalify _NO_

Basic Factors

1. Firm Price _✓_
2. Negotiated _____
3. Special Equipment _____
4. Construction Time _180 CAL.DAYS_
5. Adequate Labor _OK_
6. Adequate Equipment _OK_
7. Adequate Tools _OK_
8. Site Conditions _GOOD_
9. Unusual Problems _NONE_
10. _____
11. _____
12. _____

Work Sheet

Estimate No.: _M386_

	E1	E2	E3	TOTAL
FIXT A		12		12
B			8	8
C		50		50
D		16		16
E		20		20
F		2		2
G			1	1
H		4		4
J		14		14
K			5	5
L	1			1
M				-0-
N		11		11
P		1	1	2
R	6			6
S		3	9	12
T	2			2

Telephoned Quotations

Job: RIVER HOUSE CLUB

Supplier: STAR

Person Quoting: MIKE

Estimate No.: M386

Estimator: ED

Time: 11:00 A.M.

Date: 4/1

Quantity	Description	Price	Net
1	SERVICE / MAIN PANEL		
2	FLUSH PANELS : A & B		
1	SURFACE PANEL : C		
1	T.C. 18 X 24 X 6 SURF.		
1	CONTACTOR 100/2 P	1500.00/LOT	
		TOO HIGH	
		Total	1500.00

F.O.B. Jobsite ✓

Tax Included NO

Installed NO

Plans & Specs. ✓

Addendums 1·2·3

Division 16A

Includes:

Excludes:

Telephoned Quotations

Job: RIVER HOUSE CLUB   **Estimate No.:** M386
Supplier: VALLEY   **Estimator:** ED
Person Quoting: OMAR   **Time:** 11:20 A.M.
Date: 4/1

Quantity	Description	Price	Net
1	METERING / MAIN PANEL		
1	PANEL A FLUSH		
1	" B "		
1	" C SURFACE		
1	TEL. CAB. 18X24X6 SURFACE		
1	LIGHTING CONTACTOR, SURFACE 100 & 2B	1200.00/LOT	
		Total	1200.00

F.O.B. Jobsite ✓   Includes:
Tax Included   NO
Installed   NO
Plans & Specs.   ✓
Addendums   1.2.3   Excludes:
Division   16A

Pricing Sheet

Job: RIVER HOUSE CLUB   Estimate No.: M386

Work: ELECTRICAL   Sheet: 2 of 6

Estimator: ED   Checker SUE   Date: 4/1

Description	Qty.	Price	Per	Extension	Hours	Per	Extension
FIXT. A	12	SEE	SPREAD	SHEET	.70	E	8.40
B	8				.30	E	2.40
C	50				1.00	E	50.00
D	16				.40	E	6.40
E	20				.80	E	16.00
F	2				.40	E	.80
G	1				—		4.00
H	4				1.00	E	4.00
J	14				.80	E	11.20
K	5				1.00	E	5.00
L	1				—		2.55
M	0						
N	11				.30	E	3.30
P	2				.50	E	1.00
R	6				.75	E	4.50
S	12				.25	E	3.00
T	2				.35	E	.70
Total				5677.80			123.25

Chapter 3

Bids

Many types of bids are used in construction estimating. The bidding documents specify the type of bid required. Usually the documents contain a bid form that must be used when submitting sealed bids. Contractors usually settle on one or two types of bids that suit their operation best.

Types of Bids
Many of the bids used for prime contract work are similar to those used for subcontract work.

Prime contracting generally involves the following types of bids:
- Competitive Bidding on Advertised Jobs.
- Competitive Bidding to Selected Contractors.
- Competitive Bidding to Pre-qualified Contractors.
- Competitive Bidding with Subcontractor Listing.
- Competitive Bidding with Assigned Subcontractors.
- Negotiated Work.
- Cost Plus a Fixed Fee.
- Time and Material.
- Time with Material Furnished.

- Design and Construct.
- Construction Management.

Subcontracting involves bids such as these:
- Competitive Sub-bidding on Advertised Jobs.
- Competitive Sub-bidding to Selected Contractors.
- Competitive Sub-bidding to Pre-qualified Subcontractors.
- Subcontractor Listed Jobs.
- Competitive Sub-bidding to the Owner for Assignment to a Contractor.
- Negotiated Work with a Contractor.
- Cost Plus a Fixed Fee Subcontract.
- Time and Material Subcontract.
- Time with Material Furnished by Owner, but under Subcontract.
- Design and Construct Sub-bidding.

Competitive Bidding on Advertised Jobs
Advertised jobs for competitive bid are the most common type of contracting. Most private and all public and government jobs are advertised in the local newspapers and trade dailies. Usually the owner or his representative issues a brief project summary. Sometimes a

dollar figure is listed to show the approximate anticipated bidding level. The advertisement lists the bid date, time, opening location, phone number, project name, architect, and plan deposit. As contractors arrange for plans and specs, their names appear under the advertised listing.

Competitive Bidding to Selected Contractors

The owner or the architect may want to limit the bidding on a specific project and restrict the issuance of plans and specs to a few selected contractors. These contractors, in turn, contact the subcontractors they feel could do the job at the best price.

Competitive Bidding to Pre-qualified Contractors

Owners or architects may advertise for any contractors interested in bidding on a project but require that they qualify by furnishing a statement of experience and finances. They may also want to know if a contractor has recently completed any projects similar to the one being planned. The owner or the architect then selects which contractors can bid the project.

Competitive Bidding with Subcontractor Listing

Many bidding documents require the prime contractor to furnish a list of the major subcontractors he intends to use on the project. The list is usually part of the bid form. Most subcontractors prefer this kind of bidding since it limits bid time to the submittal of the listing. Jobs that don't require a subcontractor listing give the prime contractor time to shop around for lower prices after the owner has opened the bids. This can cause problems for the owner and the subcontractor. "Bid shopping" may force a subcontractor to reduce his bid to get the contract. Then he must look for ways to recover the cut price. One way is to reduce the labor needed to complete the job. This often results in resistance to making simple changes which ordinarily would be easily taken care of.

Competitive Bidding with Assigned Subcontractors

For the major subcontractor this type of bidding is almost like prime bidding. The owner or the architect calls for bids from the major subcontractors in the same manner as for prime

contractors. This eliminates bid shopping and offers the subcontractor a fair chance for the job. After the bids are opened, the subcontractors chosen are assigned to the prime contractor.

Negotiated Work

Contracts are negotiated when an owner doesn't want to advertise for bids and is satisfied that a particular contractor will do the best job. The owner or his representative sits down with the contractor and negotiates the project contract. The contractor must have backup pricing from one to three subcontractors for the major divisions of the contract. Negotiations may include alternate materials and construction methods if costs become a problem. Many owners feel that negotiated work is the best way to get the job done.

Cost Plus a Fixed Fee

The contractor can hardly lose on this type of contract. The owner pays for all job costs and pays the contractor a fixed fee or percentage of the overall costs. Sometimes this is the best way to get an early start on the job while the design team is still working on the plans. When time is critical, this method is the one to use.

Time and Material

Existing conditions often make a job difficult to plan. Many remodeling projects are like this. If the owner knows what the results must be, a time and material job may be the best way to go.

Time with Material Furnished

If the owner has a ready source for most of the material to be used on the job, he may want to contract for the labor only. This is usually the case when the owner has most of the material on hand but does not have the skilled labor needed to construct the job. This can be risky for the contractor because labor is the most difficult part of the job to predict. Make sure the contract includes a contingency clause to offset the higher risk.

Design and Construct

Under this contract the owner calls for bids from contractors who can design and also construct the job. The owner issues a written specification of needs to the contractors who then prepare a specific design and estimate for the project. Contractors are generally selected for their expertise in a particular field.

Construction Management

Larger projects requiring a great deal of coordination often use construction management contracts. Some contractors specialize in construction management, and subcontract the actual building of the project to another contractor. The "C.M." contractor usually lines up the other trade subcontractors.

Integrity

As estimator you are the key to your company's success. If you're careful, accurate, thorough and honest, your firm will very likely be busy and prosperous. Your dealings with others reflect the integrity of your organization. Being fair and professional ensures a good reputation for both you and your company. The estimator who violates the trust and confidence of those he deals with won't survive in the competitive world of construction contracting. Contractors don't want estimators who are unethical and unreliable. Customers won't tolerate them, either.

Be decisive in your work, yet try to keep an open mind. There are always new developments in products and installation procedures. Many of these can improve your estimating techniques. Some schools offer courses where estimators can sharpen their estimating skills.

Bid the type of work your company excels in. Stepping outside of your company's field can lead to problems. And it's just as dangerous to take on too much work. As an estimator, you help control company workload and can avoid the headaches that usually come when contractors try to grow too fast.

Bonding

The general conditions of a contract specify if the contractor is to furnish bonding for the job. Most medium to large jobs require at least a ten percent bid bond or that a cashier's check be submitted with the bid form. The bid bond is to guarantee to the owner that the contractor will enter into contract for the bid price if the bid is accepted. Bid bonds submitted by unsuccessful bidders are returned after the contract is awarded to the successful bidder.

The general conditions usually require a *performance bond* when the contract is signed. The performance bond guarantees to the owner that the contract will be completed for the bid price should the contractor withdraw from the project.

Most contractors are limited to the dollar amount they can contract for. To guarantee the contractor's performance, the bonding company must know something about the contractor and the proposed project. The contractor usually must furnish a current financial statement and an accounting of completed projects. After careful review, the bonding company may agree to limited bonding on a trial basis. They will increase the contractor's bonding capacity when he has completed several profitable jobs

Even after bonding has been arranged, don't bid the job until you know your company has the financial capability, manpower and skills needed to do the job. Consider carefully the amount of uncompleted work on your contractor's books.

When a performance bond is required, give the bonding company all the information they require and allow them ample time for review. Identify the price range of the jobs you plan to bid.

Performance bonding helps you because it provides built-in protection. Most bonding companies have years of experience with construction bonding. They know when a contractor is dangerously extended. They won't be reluctant to tell you if you're about to plunge in over your head.

The general contractor might require that the subcontractor have bonding to cover his portion of the bid. Usually the sub will be asked at bid time if he can obtain a performance bond. If the subcontractor indicates that a bond can be furnished, bids the job, and later reneges after the award of the general contract, he can be held liable for any loss to the contractor.

Chapter 4

Job Walk

The job walk is an inspection of the job site. Plan to make a job walk before starting the detailed take-off. Most contracts require that the contractor visit the site to become familiar with the actual site conditions. This is to hold the contractor responsible for covering any additional costs the site conditions may cause. An ideal building site is shown in Figure 4-1. You won't find many as good as Figure 4-1 and many will be much worse.

The site plans usually show where the improvements are to be made and identify key landmarks on the site. *Benchmarks* show the boundaries or limits of the work. In many cases the architect or designers have not visited the site themselves before preparing the contract plans. Instead they've used *as-built* drawings from a previous contract or maps of the general area. A job walk could show that the project has problems that make it undesirable from your standpoint and make a take-off a waste of time.

Conditions vary from one job site to another. When actual conditions differ from those shown on the plans, make notes so that adjustments can be made to the take-off.

In some cases a second job walk might be scheduled after the take-off has been completed. This is often done to verify certain plan interpretations against actual site conditions.

Site Conditions

Each project has its own special site conditions, so study the job carefully. Certain conditions could affect the success of the job. They can either add to or reduce the amount of work involved. Here's a list of items to look for:

1. The degree of difficulty for trenching and excavating.

2. Points of connection for utilities.

3. Points of connection for temporary utilities.

4. Access to the site and on the site.

5. Security problems.

6. Physical locations of landmarks or benchmarks that were given on the plans.

7. Distance from supply points that could cause delivery problems.

8. Distance when mileage reimbursements are required for workmen.

9. Security restrictions on military, public or private property.

10. Location and condition of existing improvements that will directly affect the new

A pre-bid job walk should always be made at the site. The picture shows an ideal building site with easy access, level lot, existing utilities and probably plenty of storage room.

Figure 4-1

work, such as: a. Pavement cutting and repairing. b. Extension of existing structures. c. Limits and the effects of demolition of existing improvements.

11. The manufacturer of existing equipment and materials that require match-up with new items.

12. Hours of the day when the existing facility will be open for the new work.

13. Special requirements to ensure public safety.

14. Special requirements for hazardous areas where inflammable liquids or materials are stored or handled.

15. Overhead obstructions.

16. Salvage items such as precious metals, equipment and other materials.

17. Removal or disposal of unusable excavation materials.

18. Conditions in structures, such as: a. Ceiling heights. b. Structural design and materials. c. Interference with other facility improvements such as boilers, air handling equipment, elevator shafts, plenums, special wall coverings, painting, floor coverings. d. Crawl spaces, pipe shafts and chases. e. Clear-

ances around machinery, equipment, stock and products. f. Safety precautions in hazardous production areas. g. Accessibility for new equipment installation. h. Hoisting and shoring requirements.

19. Existing unloading facilities that could be used by the contractor.

20. Safe staging areas.

Usually a job walk takes only a few minutes or hours. But it's time well spent. Generally, the bigger the job, the longer and more important the job walk.

Pre-bid Conference

The instructions to the bidders often include a special pre-bid conference with the owner or the owner's representative. A job walk and some take-off work will help you prepare for such a conference. You can learn a lot from the questions, answers and comments made by others taking part in the meeting.

All bidders are asked to attend. Usually the architect or the designer will be available to answer questions. Sometimes subcontractors and suppliers are invited to attend also.

The architect will study the suggestions

made at the conference. If they're valid and changes need to be made to the bidding documents, he may issue an addendum.

The architect can learn a great deal from these meetings. They help him make the job better and can help him plan future projects with similar conditions.

The owner can also benefit from the pre-bid conference. Sometimes cost savings in alternate materials or installation methods are discovered, and changes can be made before the contract is awarded. Usually it's cheaper to issue changes or addenda before the contract is awarded than after.

Contractors can ask for interpretation or clarification of certain contract requirements. They may point out additional considerations to the owner or the architect. The following is a list of topics commonly discussed in pre-bid conferences:

1. Use of suitable alternates for certain materials and equipment that may restrict competitive bidding among the suppliers.

2. Long term delivery of special material or equipment that may conflict with the completion date.

3. Responsibility for special fees, permits or assessments not covered in the bidding documents.

4. Clarification of ambiguities in the specifications or on the drawings.

5. Additional inspection authority by the owner or by the owner's representative that is mentioned vaguely in the bidding documents.

6. Substitute material or equipment.

7. Payment for storage of material or equipment, either on or off site.

8. Glaring conflicts between the specifications and the plans.

9. Value engineering. This lets the contractor share in a cost-saving idea and can be very profitable to both parties.

10. Reduced payment retention percentage as the project advances satisfactorily.

11. Bonuses for completion ahead of schedule.

12. Use of existing facilities and utilities.

Chapter 5
Estimating Aids and Tools

As an electrical estimator, you may have the use of many estimating aids and tools. These range from a hand calculator to complex computer programs. You can estimate most jobs with no more than a few pencils, paper, a rule and a map measure.

Record Keeping
Set up an estimating log to record each take-off. Use a numbering system to identify each estimate. Your completed estimates can also be useful as references for future projects. The following is a sample log:

Est. No.	Job Name	Bid Date	Contract
S-24	Boiler Hookup, City Library	7/20/26	8/15/26
S-25	Garage Wiring, Mary Ball	8/2/43	
S-26	Parking Ltg., Geana's Shop	10/2/48	
M-13	Larson Residence	3/22/53	4/30/53
M-14	Brennen Residence	4/12/53	
M-15	Jacobs Residence	4/15/53	
M-16	Sarviel Duplex	4/18/53	

Est. No.	Job Name	Bid Date	Contract
L-7	Craftsman Book Co.	8/8/65	9/1/65
L-8	Farmer's Market	10/31/72	
L-9	Marks Apartment	1/20/77	
M-17	Helen's Bakery	3/12/82	3/20/82
M-18	Spaghetti Factory	7/7/85	
L-10	Wes Towers	9/21/99	
S-27	Beth's Indian Shop	9/30/99	

S = Small Jobs, M = Medium Jobs and L = Large Jobs.

When a proposed project has been selected and recorded in the estimate log, mark the index tab of a standard file folder with the estimate number, the bid date and the time of the bid. For example:

M-16 2:00PM	Sarviel Duplex 4/18/53

On the inside cover of the folder, list all bid requests you've received from general contractors. Include the date of the request, the contractor's name and phone number, the name of the person who got in touch with you, and your bid price for that job. Note the differences in the prices you've given to each contractor. These differences are probably due to the extent of the work required for each job. But they may point out contractors you've had problems with. If a contractor has been troublesome on a job, your next bid to him will probably be higher.

Write the estimate number on the bid information sheet and place it in the folder. Do the same for all notes from the job walk and the pre-bid conference. This job information is important and should be available to you or anyone who takes over for you in your absence.

Paper and Pencils

Begin the material take-off after reading the bid documents. Every sentence in the specs and every mark on the plans can affect your cost. Your company is charged with following the bid documents precisely, so you have to know what they require. Don't assume you've seen these specs before. You probably have seen them or a variation of them many times, but a single sentence can increase all of your costs by several percent. Don't miss that sentence! There is no substitute for reading the specs carefully and examining the plans in detail.

Make marginal notes as you go through the plans the first time. Underline important or unusual items on the specs. Identify portions that leave a question in your mind or need clarification. When you have completed your "first pass" over the plans and specs, it's time to begin the actual take-off.

I use a pad of standard size (8-1/2 inch by 11-inch) lined paper for taking off material quantities. Draw a one-inch margin down the left side of the lined sheet. Leave all sheets in the tablet until the material take-off has been completed and all of the listed material has been transferred to the pricing sheets. Then remove the work sheets from the tablet, staple them together, and place them in the estimate folder.

Use a soft or medium-soft pencil to list the material on the work sheets. Write as clearly as possible to reduce the risk of error when transferring data to the pricing sheets.

With colored pencils you can color code the plans as the material take-off progresses. With a colored pencil check off each item on the plans when you have listed it on your take-off sheet. The following is a simple, effective color code:

Red: Use on all counted items such as lighting fixtures, wiring devices, outlet and junction boxes, precast concrete boxes and vaults, equipment, hookups, motor control and safety switches.

Yellow: EMT conduit and fittings, flexible conduit and fittings.

Brown: Rigid conduit and fittings, PVC conduit and fittings.

Blue: All items furnished by others, such as mechanical equipment and controls and owner furnished items.

Green: All existing "to remain" items.

Orange: Grounding and bonding systems.

When you've completed the take-off by marking all of the identifiable items, your plans will be a colored picture of the electrical work to be done. When the project is underway, the plan becomes a job reference. You may even discover some serious problems early enough to request a clarification or a contract adjustment. The installers may find a better or cheaper way to make the installation. Color coding is a simple but effective way to increase take-off control.

The next chapter, Chapter 6, describes how the actual take-off is made.

Rules and Measures

Every electrical estimator needs an architect's scale or rule. Use it to scale off conduit, trenching, available equipment space, and other items shown on the drawings. Use a 12-inch triangular rule with scales from 1/16 to 1 inch. Be careful not to damage the edges of the rule. Damaged edges produce broken or uneven lines and reduce the clarity of your drawings. Most architects don't use their rule as a guide for drawing lines. That may be a good policy for you too.

Most building plans have a scale of 1/8 inch to the foot. Others use 1/4, 1/2, 3/4 or 1 inch to a distance specified on the drawing. Some plans use one scale on one sheet and another scale on another sheet. Always be sure to select the right scale. And watch for obvious errors in the plan scale. The difference between a correct scale and an incorrect one will have a big effect on the bid price.

Some estimators use a rule to check all measurable items on the drawings. Others use a small tape measure to find conduit lengths. But most estimators use a *map measure* or a *roto-meter* to take off linear measurements. These have a small wheel which you roll along the line to be measured. A scale on the instrument converts plan distance to actual linear feet. Some type of flexible measuring tool is particularly valuable when scaling off along a curved line.

A map measure resembled a stop watch. It's about 1¾ inches in diameter and 1/2 inch thick. The face is like a watch, but with numbers from 1 to 100. There are two smaller dials in the face. One has numbers from 100 to 1000; the other has numbers from 1000 to 10,000. A reset button at the top of the instrument sets all three dial pointers to zero. A small wheel protruding from the bottom activates the dials when the instrument is moved across a drawing.

The main pointer shows the distance the small wheel has moved. When the large pointer moves completely around the instrument face, a smaller dial will point to 100. When the smaller dial's pointer moves completely around its face, the next dial will point to 1000.

The map measure is a delicate instrument. Keep it in a box when not in use. And keep it clean and in good condition. When you have a variety of map measures, mark the boxes to indicate the scale.

Some estimators use a roto-meter for linear measurements. The standard scale is 1" = 10'. The estimator must convert each measurement to the plan scale to determine the correct distance. Most roto-meters do not have a reset button. You must move the roto-meter in reverse to reset the dial.

Counters

Counting devices are commonly used for counting items such as lighting fixtures, switches, and outlets. They provide accuracy and help eliminate confusion during take-off counts. The counter I use is an inexpensive hand device about 1¾ inch in diameter and 1⅛ inch thick. It's activated by a push button, and registers from 1 to 9999. A reset knob sets the indicator back to zero.

Electronic measuring devices are available that can be easily adapted to electrical estimating. They have a probe about the size of a pencil. You insert colored lead to mark the drawings as items are measured. As each item is marked, the probe transmits an impulse to a deck where indicators give a total as the work progresses. You can select a different indicator on the deck for each item, then return to a previous indicator. By attaching a colored wheel to the probe, an indicator will total on a selected plan scale. The probe leaves a color trace on the drawings to show that the item has been taken off. Usually more than one deck is used during the take-off. If you do take-off work all day and every day, a tool like this will be a valuable aid.

The National Electrical Code

The National Electrical Code is an essential estimating aid for the electrical estimator. In most cases you have to determine the sizes of conduit, wire, cable and outlet boxes to be used in the proposed project. The code is revised

The map measures are used when taking off lengths such as conduit systems on plans. The one on the left is for a 1/8" scale. The middle one is for 1/4" scale and the one on the right is for 1" scale. The other device is a counter that can be used on large quantity items such as lighting fixtures.

Figure 5-1

every three years, so be sure to have a current copy. Most good technical bookstores have a copy or can get one for you.

The code has many tables and examples for quick reference. The following are some tables commonly used by many estimators:

Table 310-16, Allowable ampacities of insulated conductors.

Table 310-17, Same as above, but in free air.

Table 310-18, Same as -17, but 110 to 250 degrees C.

Table 310-19, Same as -18, but in free air.

Table 3A, Maximum number of conductors in conduit or tubing.

Table 3B, Same as above, but continuation.

Table 3C, Same as above, but continuation.

Table 370-6a, Metal boxes, maximum number of conductors.

These code tables are a valuable source of information for all electrical estimators. You need a working knowledge of the current code in force where the work is being done. Many local authorities enforce additional rules that may be more stringent than the NEC, and most design engineers specify additional contract requirements to ensure a high quality job. That's because the NEC is considered to be the minimum requirements.

If your bookstore doesn't have a copy, the National Electrical Code is available from the National Fire Protection Association, Attention: Publication Sales Dept., 470 Atlantic Avenue, Boston, Massachusetts, 02210.

Some suppliers, manufacturers and trade associations offer the current code books to their members and customers at little or no cost. Usually the books are available immediately upon code issue, so be sure to request the book as early as possible. If a charge is made, it's usually to cover handling—about a dollar or two.

Adding Machines and Calculators

Adding machines and calculators are extremely handy estimating tools. Every experienced estimator has his own preference. Some prefer an adding machine with a tape printout. Certain calculators are also available with a tape. Look for a machine that is easy to use, sturdy, fits your hand, is easily serviceable, and which produces a clear summary.

Computers

Once, computers were too expensive to be used as an estimating tool. Only large contractors could afford to buy, program and use a computer. Today, there are many inexpensive computers on the market.

Microcomputers are the smallest computers. Many models are available for various applications. *Minicomputers* are larger, more versatile, and more costly. They require expensive programming and systems work to be worth the investment.

The computer for you, if any, depends on the size of your operation and the amount of time and money you are willing to spend to get the programs running. Nearly all computers are good. Some may be better than others. The real difference is in the programming. Don't even consider a computer until you have seen the program you need work on that computer and have actually found that the program fits your needs. Many programs will work on only one model of computer. So find the program first and then buy whatever computer is needed to run that program.

As of this writing I have not seen a really professional quality electrical estimating package suitable for most small electrical contractors that will run on the low-cost microcomputers. That will change. But find satisfied users and try using the programs yourself before you buy any computer.

Regardless of the machine you choose, the key to the system is the programming. Developing a good estimating program can take hundreds or thousands of hours. And skilled programmers charge more per hour than your tradesmen cost in the field. Don't assume that you can write your own programs or have exactly the program you need written in a few days or weeks.

Computers are an excellent tool for managing daily business activities. They can be programmed for bookkeeping, payroll, general ledger, material inventory, equipment records, and many other general business needs. With a computer you can process, store, and access information quickly, efficiently, and easily.

But a computer is only a tool — like another pickup truck. Do without it if you can and buy one only when the need is obvious to everyone. It complements your estimating practices and procedures—it does not replace them. You still must evaluate each job and properly assemble your take-off data. No computer can correct a sloppy or inaccurate take-off.

Chapter 6

Systematic Take-off

Every estimator should develop a systematic approach to taking off a job. This chapter explains a workable method that can be applied to most electrical construction bidding where a *lump sum price* is required. The system can be easily adapted to other bidding requirements such as *bid alternates, unit pricing* and even *change orders*. No matter how large or small the job, identifying every cost item is the key in predicting the correct price.

Job Selection
Knowing what jobs to bid is very important. Check the plans and specifications of the jobs you are considering. These are available from the general contractors, awarding authorities, architects or owners. Don't bother to bid work that you don't have the tradesmen, financial resources, managerial talent or technical ability to handle. Taking the wrong jobs or taking too many jobs can be fatal for any electrical contractor. Look for projects that offer a good chance for successful completion — projects like what you have handled before and done well on.

Scope Of Work
Read the bid instructions, the complete bid form, the proposed contract, and the general and special conditions. Note the completion date and the amount of liquidated damages, if any.

Usually the general contractor gets only a limited number of sets of project plans and specifications from the owner or the architect. To get the most subcontract bids, the general contractor will split most of the sets of plans and specifications into trade groups (civil, architectural, structural, plumbing, air conditioning and heating, electrical, and landscaping) and will only issue one section to a prospective subcontractor for preparing a bid. This way the general contractor can make the fewest sets of plans and specifications available to the largest number of subcontractors during the bid period.

Generally the contractor will issue the trade package to each subcontractor for only a few days. But he'll keep a complete set of the bidding documents on hand for examination by the subcontractors during the bidding period. All addenda or changes to the bidding documents are noted in the complete set.

It's your responsibility to understand the project requirements and to include on your set of the plans all addenda or changes issued during the bid period. Never depend on the general contractor to notify you of the addenda or changes. Check with him occasionally to be sure that your bid will be complete when the bids are called for.

When you bid a section of the job, the general contractor will ask if your bid covers certain changes or certain hookups or wiring listed in other sections of the bid package. You are expected to bid your section as it is specified. If you find that certain items are questionable and would *not* normally be covered under your section, be prepared to *exclude* those items from your bid. If you find something that is questionable and *might* be covered under your section, be prepared to *include* that item in your bid. Always try to state your bid as accurately as possible. Stay abreast of addenda or changes made during the bid period. There will nearly always be a few on any significant project.

Many general contractors want a written statement of what your bid will include or exclude before the bid time. This is to get a better idea of the scope of work covered in your bid. The contractor may find certain items in your bid that can be performed by other trades, thus lowering the overall bid price. Such items might include trenching, placing concrete pads and vaults, and painting. The contractor might find that he can save money by doing all of the trenching for the plumber, electrician and landscaper.

Be sure to read all instructions, addenda and changes when checking the full set of bidding documents prior to bid time. It takes experience and study to fully understand addenda and changes. They are usually brief and may just change a word or two in the specifications or add a note on a drawing. But these little changes can make a big difference in the scope of work, and that will directly affect your cost. Any note can change the scope of your work. Take notes on or underline key items that will affect your bid.

Study the bid form that you will use to present the bid price to the owner. Sometimes the bid form is changed to include additional pricing categories, alternates, breakout pricing for accounting, or changes in the bid date or time. Any change in the bidding form or in any of the bid documents can affect the bid price.

Be absolutely sure of the work you are bidding on. If you find that the bidding

documents have been changed during the bid period and you have not reviewed the changes, pass on the job rather than bid it "blind". Sometimes the changes are minor and the general contractor can explain each change. You can adjust your bid with reasonable certainty. But this is risky and could result in a bad bid.

Check The Bid Package
Check the plan index to be sure you have all of the plans in the set. Make sure the plans are readable. Do the same in checking the specifications. And make sure you understand the scope of work explained in the general requirements.

Plan Take-off
After the scope of work has been determined and the bid procedure is clear to you, prepare the job folder as described in the last chapter. Assign a bid number to this job to make identification easier for everyone in your office.

The next step is to complete the material take-off as explained in Chapter 5. The listing of materials isn't difficult if you work carefully and systematically. Develop some system of checking off each material as it's measured or counted. Follow that system on every bid you make.

When all materials are listed, transfer only total quantities to your pricing sheets. Collect groups of similar or associated materials together on these pricing sheets. Again, this makes forgetting something less likely. List quotes from suppliers or electrical wholesalers first so they'll be easy to find later. Your categories on the pricing sheets probably include the following:

• Service and distribution equipment/panels/transformers.
• Lighting fixtures and accessories.
• Lamps.
• Conduit and fittings.
• Boxes and box fittings.
• Wire and cables, splices, terminations.
• Wiring devices and cover plates.
• Safety switches, motor control, gutter, pull boxes.
• Hookup equipment.
• Trenching, excavating, concrete, core drilling, light bases.
• Demolition.

Expand this list to include as many more items as needed. But don't list work that will be subcontracted to specialty contractors: communications systems, burglar alarm systems, TV systems, telephone systems and other specialty installations. These should be listed after all taxable items have been tabulated and the tax calculated and added in. You have to pay tax on materials you buy but not on materials supplied by subcontractors.

Add subcontracted work on the *end sheet,* or *bid summary,* which serves as a cover sheet for the estimate package. The bid summary also serves as an index to the other sheets and sections which make up your estimate.

Work Sheet

Use a separate work sheet for listing the lighting fixtures. List the fixture types in the left margin on the work sheet, using the lighting fixture schedule on the plans or in the specifications. Don't forget to identify the work sheet with the estimate number.

At the top right of the work sheet indicate the drawing number if more than one drawing is used to show the lighting systems. As the lighting fixtures are counted, list the quantities on the work sheet opposite the corresponding fixture type. When all of the lighting fixtures on that drawing have been counted and listed, start another column for the next drawing and repeat the counting and listing process. A large job may have several lighting plans.

You will use this lighting worksheet to locate the various fixture types as the estimate proceeds or when the work is actually done. For example, suppose after counting the fixtures you find a note requiring a certain fixture accessory in certain rooms. Such an accessory might be a radio suppressor, dimming ballast or a suspension device. The work sheet can guide you to the page where that fixture is being used. Sometimes a supplier will offer fixtures in tandem channels at a cost saving. The lighting worksheet will help you find the fixtures that can be installed in tandem.

Pricing

There are many ways to price materials. Most jobs are priced in units costs. Usually fixtures are priced separately. The electrical wholesale distributor usually has a lighting fixture department where salespeople search the manufacturers' catalogs for the listed lighting fixtures. They may try to find an alternate manufacturer that can offer an "equal" fixture at a lower cost. Most electrical contractors depend on the electrical wholesale distributor for gathering and sorting fixture prices. In many cases, the distributor has access to a special take-off service and receives fixture take-offs on many jobs that are being bid. If they don't have access to the fixture take-off for the job you're bidding, they may want you to supply the take-off. They may even send a salesperson to your office to get the listing. It's very important to keep the distributor up to date on the jobs you're planning to bid.

Sometimes the distributor will price each type of fixture separately. They may offer alternative fixtures at lower cost. This makes *you* responsible for deciding whether to take a chance on the alternate in the hope that the owner will approve the item. Or, the distributor may offer a lot price for *all* the lighting fixtures for the project. This makes *them* responsible for project compliance and for the correct number of fixtures required. If the alternates are rejected by the owner, a resubmittal or a corrected submittal of the item must be made. Usually there is a cost difference involved, so be sure to have your notes in order. But try to avoid resubmittals. They usually require a lot of extra time. And some jobs have too tight a schedule to permit resubmittal.

One reason why distributors offer alternate fixtures is that they don't have access to all lighting fixture manufacturers. To bid a job, the distributor may have to find an alternate that has a good chance of being approved. Most designers recognize this and accept certain alternates, provided they don't lower the standards set for the job.

Often, the alternate being offered is a lower quality fixture at a much lower cost. Before gambling, decide who will cover the cost difference for the specified fixture. If your firm stands to be tagged with the added cost, you might want to add a portion of that difference in your bid.

Sometimes the specified manufacturer will feel that the owner will not accept any other fixture manufacturer and may raise his prices to take advantage of the situation. An alternate price then becomes even more attractive. If the difference is large enough, an offer to use the alternate fixture can be made after the contract has been awarded. It's good practice to share the savings with the owner.

Every electrical wholesaler knows that submitting prices too early is dangerous. Prices offered too early before bid time invite undercutting by competitors. Thus, distributors are reluctant to release their best prices early. Sometimes their early prices are left a little high so a more competitive price can be offered just before bid time. This tends to waste your time because you've already extended the higher costs to find the job total. You're going to be too busy at bid time to recompile your figures again. Ask your material supplier to give you his best price the first time.

Developing the trust of those you are doing business with is extremely important. When the distributor finds that your word is good and that you can be trusted with the best pricing, he will provide you with the best service he can offer. Don't peddle prices from one distributor to another. None will trust an estimator who plays that game.

Here's how to price lighting fixtures:

Unit cost: Priced individually for each type.

Lump sum: Priced in a lot price for all of the fixtures.

Combination: Unit cost and lump sum. Selected fixtures by unit cost and one or more types at a lot price.

Alternate: Priced by unit cost.

Alternate: Priced in a lot price for all of the fixtures.

Alternate: Unit cost and lump sum combination.

Accessories: Plaster frames, stem and canopies, tandem housings, connectors/splices/couplings, special slope adapters, remote parts, radio suppressors, dimming features, in-line fuses and a host of other special requirements.

Accessories can be priced separately, included in the lot price, or priced both ways to match the fixture pricing. Frequently, the accessories are not mentioned at all and will be furnished by your firm to complete the job. The supplier's price does not cover accessories unless specifically stated at the time the prices are offered.

Here's how to keep track of all lighting fixture prices and to be ready for alternate and accessory pricing. Start a work sheet with room for prices from all expected electrical wholesale distributors on all of the project's lighting fixtures. An easy way to do this is to set up a sheet similar to the take-off work sheet. The sheet can be prepared immediately after the take-off has been completed and all of the fixture details have been studied and listed. This work sheet is called a *spread sheet*.

When the list of fixtures and accessories extends to more than one spread sheet, attach the sheets with tape to form one long spread sheet. It's important to keep all of the pricing information on a single spread sheet. The estimator then has all of the fixture information in one place for easy reference. This allows quick comparison with the price lists to find the best possible price. And it's easy to detect a mistake in the pricing extensions when all are on a single spread sheet.

Chapter 7

Residential Take-off

The worksheets and price sheets in this chapter are for the residence shown in Figure 7-1. The drawing shows an electrical layout to be installed under a subcontract awarded by a general contractor.

All interior and exterior walls are wood frame. The inside walls will be covered with 5/8-inch gypsum drywall, except in the garage, which will be 3/4-inch drywall. The wall exteriors will have 1/2-inch siding. There is a full attic, and a crawl space under the house. The garage floor is concrete on grade.

The project is to take 180 calendar days, with a starting date in mid-spring. There is a penalty clause for damages should the job not be completed on time. The penalty is referred to as *liquidated damages* and amounts to $100.00 for each day of delay. Allowances will be made for delays beyond the control of the contractor such as rain, strikes and owner requested changes.

The job walk reveals that the lot is reasonably level and is located in a small, new residential development under construction within city limits. There is no electric power available for construction; the general contractor will furnish an electric generator for use by all trades on the job. The soil looks soft to

medium firm, with a little gravel mixed in. Trenching should be no problem. The project has the advantage of a watchman at night, little traffic and few children in the area.

Use the forms in the back of this book to make an estimate of costs for the electrical work shown on the drawings. Run off several copies of each form on a copy machine for the forms you need for this exercise. An "instant printer" can give you better quality reproductions when you need a larger quantity of forms.

Study the drawings and become familiar with the symbols. Most symbols are industry standards, but be alert for less common symbols that may not mean the same thing to everyone. Some designers deviate from the standard symbols or invent new symbols. Never trust the symbol list to be complete; it usually shows only the most common devices. A list of common electrical blueprint symbols is on page 100.

Start the material take-off by listing the lighting fixtures as shown on the sample work sheet (Figure 7-2). After all of the lighting fixtures have been counted and listed, count and list the wiring devices. Check the work sheets in Figures 7-3 and 7-4 to see how other materials are listed.

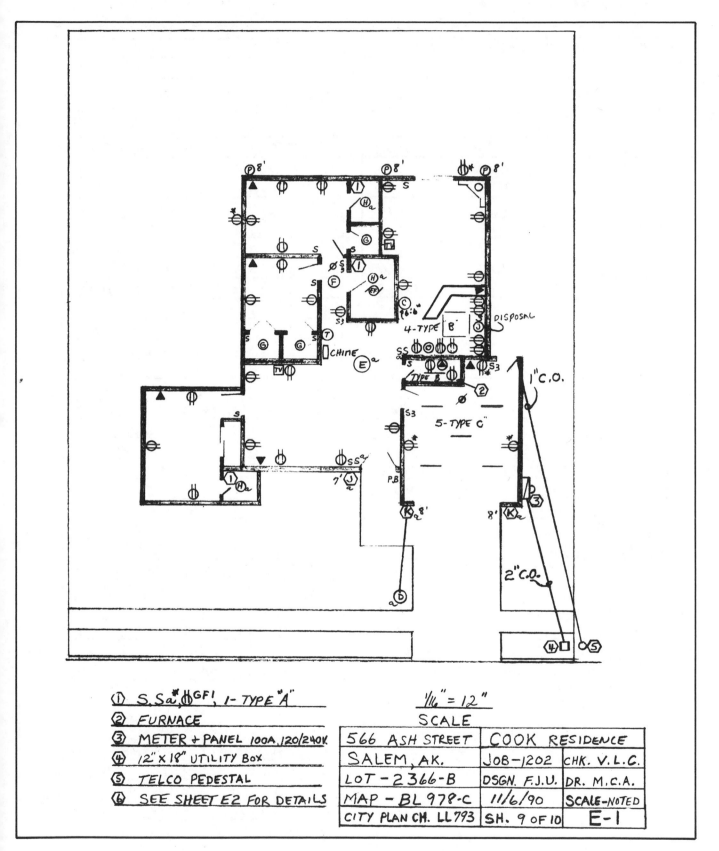

① S.Sa.ʰGFI, 1- TYPE "A"
② FURNACE
③ METER + PANEL 100A,120/240V
④ 12"x18" UTILITY BOX
⑤ TELCO PEDESTAL
⑥ SEE SHEET E2 FOR DETAILS

¹/₁₆ " = 12"
SCALE

566 ASH STREET	COOK RESIDENCE	
SALEM, AK.	JOB-1202	CHK. V.L.C.
LOT-2366-B	DSGN. F.J.U.	DR. M.C.A.
MAP-BL978-C	11/6/90	SCALE-NOTED
CITY PLAN CH. LL793	SH. 9 OF 10	E-1

Figure 7-1
Plans for Sample Take-off Residence

LIGHTING FIXTURE SCHEDULE		
A	MARCO B113	F40CW
B	LITHONIA C240	2-F40CW
C	LITHONIA UN240 WG9	2-F40CW
D	PRESCOLITE 5812 FP	60 IF
E	CHANDELIER BY OWNER	—
F	PRESCOLITE 7810	2-60 IF
G	LEVITON 9875-2	60 IF
H	MARCO J322-R40A	250R40/1
J	PRESCOLITE 5942-4	60 IF
K	PRESCOLITE 5940	100 IF
P	PRESCOLITE 5322	75 PAR38FL

SYMBOLS	
S	LEVITON 1451-1, SMOOTH IVORY PLATE
S3	LEVITON 1453-1, SMOOTH IVORY PLATE
⊖	LEVITON 5248-1, SMOOTH IVORY PLATE
⊖	LEVITON 279 , BROWN OR BRASS PLATE
◀	LEVITON 5207, BROWN OR BRASS PLATE
©	LEVITON 628-1
TV	LEVITON 86018
▶	LEVITON 86018
P.B.	LEVITON 9701-1K
⬚	CHIME - BY OWNER
⊘	NUTONE S231 SMOKE DETECTOR
EF	EXHAUST FAN - BY OTHERS
O	TELCO PEDESTAL, CONFIRM LOCATION
⬜	UTILITY BOX, CONFIRM LOCATION

NOTES	
1.	FIXTURES AS LISTED, OR EQUAL
2.	WIRING DEVICES AS LISTED, OR EQUAL
3.	ALL WIRE - COPPER
4.	UNDERGOUND CONDUITS - 24"
5.	GROUNDING - NEC & LOCAL CODES
6.	INSTALLATION - NEC & LOCAL CODES
7.	UNDERGROUND - UTILITY CO. APPROVED
8.	METER/PANEL - UTILITY CO. APPROVED
9.	POST LIGHT BASE - 12"×12"×24" FLUSH
10.	FURNISH PIGTAILS - DRYER & RANGE
11.	MOUNTING HEIGHT - DUPLEX 12" CENTER
	ABOVE KIT. COUNTER
	30" IN LAUNDRY
	30" IN GARAGE
	24" OUTSIDE
	30" AT FURNACE
12.	MOUNTING HEIGHT - SWITCHES 48" CENTER
	UNLESS DIRECTED TO CLEAR OBSTACLES.
13.	TELCO & TV OUTLETS AT 12" CENTER
	ABOVE KIT. COUNTER
	30" IN GARAGE
14.	CHANDELIER BACKING FOR 300 POUNDS
15.	KITCHEN LIGHTING IN DROP CEILING
16.	DROP CEILING BY OTHERS
17.	CONNECT ALL EQUIP. FURNISHED BY OTHERS
18.	FURNISH CATALOG DATA ON ALL SUBSTITUTES
19.	BALANCE PHASES AT PANEL
20.	IDENTIFY EACH PANEL CIRCUIT BREAKER
21.	TEST ALL CIRCUITS
22.	FURNISH ELEC. PERMITS & INSPECTION CARD

566 ASH STREET	COOK RESIDENCE	
SALEM, AK.	JOB-1202	CHK. V.L.C.
LOT - 2366-B	DSGN. F.J.U	DR. M.C.A.
MAP-BL 978-C	11/6/90	SCALE-NONE
CITY PLAN, CK. LL 793	SH. 10 of 10	E-2

Figure 7-1 (continued)
Plans for Sample Take-off Residence

Work Sheet

Estimate No.: _S 115_

	E1
TYPE A	3
B	5
C	5
D	1
E	1
F	1
G	3
H	3
J	1
K	2
P	3
S	9
SS	2
S⊖GFI	3
S3	4
⊖	30
⊖⊖	2
⊖GFI	5
⊖	1
◑	1
©	1
⊘SD	2
PB	1
◀	6
TV	2
DISPOSAL	1
CHIME	1
Ⓣ	1
ⒺⒻ	1
METER	1
⊙ OVEN	1

Figure 7-2

Work Sheet

Estimate No.: S 115

	E1
4/0 BOX	23
1G	54
2G	4
3G	3
4/S BLANK	1
1/2" PVC	20
1"	60
2"	30
1/2" FA	2
2"	1
1/2" Elbow	2
1"	2
2"	2
1/2" COUP.	1
1"	1
2"	1
2" RIGID	5
2" L/N	2
2" INS GR BUSH	1
UTIL. BOX	1
TEL-PED	1
TRENCH	80
D BASE	1
E BACKING	1

Figure 7-3

Work Sheet

Estimate No.: _S 115_

	E1	
2/C #14	1140	
3/C #14	300	
2/C #12	250	
3/C #12	260	
3/C #10	70	
3/C # 6	70	
3/C #18	100	
TV CABLE	100	
1/2" CONN	12	
3/4	2	
#12 TW	120	
#6 GR.	50	
3/4 GR. CLAMP	1	
STAPLES	100	
1/2" FLEX	10	
1/2" CONN	4	
PullSTRING	100	
1/2" PENNIES	2	
1"	2	
2"	2	

Figure 7-4

When all material is listed and all items on the drawings have been marked to indicate that they have been counted, you've completed the first step of the estimate. Remember that what you have counted and listed is the very *minimum* amount of material that will be used in the project. The list does not consider changes in the wiring plan because of obstructions created by other trades. It doesn't include any waste by the field workmen installing the material, nor is there any way to figure lost, damaged or defective material.

The most important thing to remember at this point is to be sure that all of the material has been accurately counted and recorded. If you have any doubts, do a recount or a remeasure.

As the material is transferred to the pricing sheets (Figure 7-5), certain items can be clarified. When EMT conduit has been listed on the work sheets and the connectors have been counted, the size and quantity of couplings can be calculated and listed. The same is true of the conduit straps: they too can be calculated from the size and footage of the conduit.

After the NM (non-metallic) wire has been measured, the number of staples and the number of holes to be drilled can be calculated.

These are all items that require labor and must be listed.

Outlet boxes will need covers. There are many kinds. Blank covers, single gang switch rings, two gang switch rings, handy boxes in various depths, octagon boxes (some with plaster rings), multiple ganged switch boxes and switch rings, bar hangers, bracket boxes, wood backing, bracing, surface junction boxes, and flush junction boxes are examples. They all should be listed so that the material price and the labor can be applied.

Each item on the work sheet will usually require many minor items to make the installation. Now is the time to put them all down.

A good, thorough take-off done in an orderly manner simplifies preparation of the material submittal lists. These lists show the types of material to be used for the job. The take-off can be used when purchasing most of the material. It can also be used when coding the job for accounting purposes, labor control, tool and equipment control, and scheduling.

Your material take-off is the foundation on which the whole estimate is built. A mistake here will invalidate every subsequent calculation. Take the time to do an accurate material take-off.

Pricing Sheet

Job: __COOK RESIDENCE__   Estimate No.: _S-115_
Work: __ELECTRICAL__   Sheet: _1_ of _4_
Estimator: _ED_   Checker _SUE_   Date: _12/15_

Description	Qty.	Price	Per	Extension	Hours	Per	Extension
METER / PANEL	1			166.00			
SURF. MTD.							
BOTTOM FEED							
120/240 V 1∅ 3W							
100A 2P MAIN							
5 15A 1P							
5 20A 1P							
1 30A 2P							
1 50A 2P							
FIXT. A	3	48.50	E	145.50			
B	5	19.20	E	96.00			
C	5	23.31	E	116.55			
D	1	—		77.00			
E	1	—		F30			
F	1	—		16.65			
G	3	1.00	E	3.00			
H	3	12.25	E	36.75			
J	1	—		60.15			
K	2	10.50	E	21.00			
P	3	3.50	E	10.50			
LAMPS 60A	5	53.00	C	2.65			
100A	2	63.00	C	1.26			
75 PAR 38/F1	3	345.00	C	10.35			
150 " "	2	351.00	C	7.02			
250 R 40/1	3	325.00	C	9.75			
F40CW	23	149.00	C	34.27			
Total				814.40			

Figure 7-5

Pricing Sheet

Job: _____ Estimate No.: _S-115_

Work: _____ Sheet: _2_ of _4_

Estimator: _____ Checker _____ Date: _____

Description	Qty.	Price	Per	Extension	Hours	Per	Extension
1/2" PVC 40	20	9.98	C	2.00			
1"	60	19.78	C	11.87			
2"	30	43.74	C	13.12			
1/2" FA	2	.18	E	.36			
2"	1	—		.80			
1/2" ELBOW	2	.46	E	.92			
1"	2	.80	E	1.60			
2"	2	2.22	E	4.44			
1/2" COUPLING	1	—		.15			
1"	1	—		.28			
2"	1	—		.70			
PVC CEMENT	1 QT.	—		6.00			
2" RIGID CONDUIT	5'	126.60	C	6.33			
2" LOCKNUTS	2	44.50	C	.89			
2" INSUL. GR. BUSH	1	—		3.27			
1/2" FLEX CONDUIT	10	17.02	C	1.70			
1/2" CONN	4	17.30	C	.69			
1/2" PUSH PENNY	2	9.00	C	.18			
1"	2	13.50	C	.27			
2"	2	54.40	C	1.09			
4/0 BOX W/BKT	23	66.15	C	15.21			
S PLASTER RING	23	45.50	C	10.47			
4/S BOX W/BKT	1	—		1.07			
S BLANK COVER	1	—		.32			
1 GANG BOX W/BKT	54	46.25	C	24.98			
2	4	84.90	C	3.40			
3	3	152.70	C	4.58			
Total				116.69			

Figure 7-5 (continued)

Pricing Sheet

Job: _____ Estimate No.: *S-115*
Work: _____ Sheet: *3* of *4*
Estimator: _____ Checker _____ Date: _____

Description	Qty.	Price	Per	Extension	Hours	Per	Extension
3/C #18 TWISTED	100'	71.00	m	7.10			
2/C #14 W/GR NM	1140	89.00	m	101.46			
3/C #14	300	152.00	m	45.60			
2/C #12	250	131.00	m	32.75			
3/C #12	260	221.00	m	57.46			
3/C #10	70	359.00	m	25.13			
3/C #6	70	1048.00	m	73.36			
TV CABLE 1PR.	100	80.60	m	8.06			
PULL STRING	100	16.10	m	1.61			
#12 TW SOL.	120	34.86	m	4.15			
#6 GR. CABLE	50	355.00	m	17.75			
STAPLES	100	5.09	C	5.09			
S PLATE 1451-1	9	152.25	C	13.70			
SS	2	305.20	C	6.10			
SS GFI	3	3515.95	C	105.48			
S3	4	200.20	C	8.01			
5248-1	30	121.80	C	36.54			
	2	251.65	C	5.03			
GFI SWR 15 F1V	5	3132.50	C	156.63			
RANGE 279	1	—		8.03			
DRYER 5207	1	—		2.93			
CLOCK 628-1	1	—		1.07			
S.D. S231	2	22.60	E	45.20			
TEL. 86018	6	280.00	C	16.80			
TV 86018	2	280.00	C	5.60			
DRYER PIGTAIL	1	—		3.76			
RANGE "	1	—		11.13			
PB 9701-1K	1	—		.78			
Total				806.31			

Figure 7-5 (continued)

Pricing Sheet

Job: _____ Estimate No.: S-115

Work: _____ Sheet: 4 of 4

Estimator: _____ Checker _____ Date: _____

Description	Qty.	Price	Per	Extension	Hours	Per	Extension
1/2" 2 SCREW CONN	12	28.40	C	3.41			
3/4" " "	2	51.80	C	1.04			
DISPOSAL HOOKUP	1	—		1.00			
THERMOSTAT	1	—		—			
EXH. FAN	1	—		1.00			
OVEN	1	—		1.50			
CHIME	1	—		FBO			
3/4" GR. CLAMP	1	—		.97			
5/8" X 8' GR. ROD	1	—		10.38			
5/8 " CLAMP	1	—		1.72			
TRENCH 10X24	50	.80	F	40.00			
" 14X24	30	1.00	F	30.00			
POST BASE	1	—		15.00			
EXCAV. " 12"X24"D	1	—		—			
UTILITY BOX 12X10X18	1	—		31.25			
EXCAV. " "	1	—		—			
LOCATE UTILITY BOX	1	—		—			
" TEL. PED	1	—		—			
DRILL STUDS	268	—		—			
BACKING - FIXT E	1	—		—			
TEST	1	—		—			
		Total		137.27			

Figure 7-5 (continued)

Chapter 8

Take-off Format

Establish a good take-off procedure that you can use on every project. Start the take-off process the same way on every job. Follow the same format when setting up the work sheets, counting and measuring materials, and listing materials on the pricing sheets. After several jobs of about the same size and type, the take-off packages will look very similar. As more take-offs are completed following the same procedure, the take-off packages still should be basically the same, even though the size of the jobs may differ.

Following a standard and a consistent procedure allows others to use the estimates you develop. Another estimator can quickly refer to your take-off to get information for another estimate. Some companies employing more than one estimator for take-off work establish a uniform format to be followed by all their estimators. Other firms have developed systems that involve other employees in various stages of the estimating process. For example, a purchasing agent may price the basic materials while someone else collects all the necessary quotations for the estimating process. The chief estimator then reviews each estimate before it is bid.

In many companies the estimator determines the project's direct estimated cost while the front office adds the overhead, profit, additive or deductive factors and follows through with the actual bid. The estimator's job ends when the direct estimated costs are established. If he's followed a clear and consistent format, those who complete the work will have little trouble understanding his take-off. Over time you will see improvements needed in the format you use, but usually these adjustments will be slight. The format may need to make adjustments for special projects.

Usually you complete the take-off and establish the basic cost of the project several days or even weeks before the job is actually bid. The plans may then be returned to the general contractor. You probably go to work on some other estimate after that and forget most details of the completed estimate. If you have to come back to that estimate weeks after you completed it, you'll be glad you followed a consistent procedure and made all calculations clear when the estimate was first made.

Eventually you'll be faced with the task of training a new estimator. When this happens, you'll want to pass on all your know-how to your pupil. If you have followed a consistent procedure, your trainee is more likely to follow the same system you have been using. That makes it easier for you to follow and monitor his work.

Change Orders and Adjustments

Most contractors and owners require an estimate and a change order before changes are made to the working plans. This submittal must list all material and labor needed for that change. Again, follow a consistent and easy-to-understand estimate format. A clear, well-presented take-off will make it easier to get approval of the change at the price you quote.

The change order format can be the same as the one used for the original estimate. Prices will fluctuate with changes in material costs. Labor will be slightly higher because of the higher cost of procurement, supervision and control, accounting, and, most of all, interference with scheduled production.

Adjust the take-off to include addendums or changes requested by the owner during the bidding. These changes can be additions or deductions. They may even be a combination of additions and deductions. A good format, with all calculations clear and easy to follow, make adjustments and changes much easier.

Recheck the Specifications

Carefully recheck the electrical specifications to determine if any additional material or labor is required. List these in the take-off. Certain tests may be required; these must be considered as a labor item.

After all material and labor items have been listed, go through the specifications once more to find the specifics on materials. Designers may list manufacturer names and catalog numbers for specific material required on the project.

Wiring devices are manufactured in many grades and are usually listed by grade and color. The specifications may list acceptable wire, such as copper or aluminum. The specified insulation may be TW, THW, THHN, THWN,

XHHW, RHW-USE, XLP, or TF. The specifications may list a certain wire to be solid or stranded, or may specify temperature ratings for the insulation.

Outlet boxes may be one-piece pressed steel or ganged. Plastic or cast metal boxes could be specified as weatherproof or suitable for dry, damp or wet locations.

Specifications usually list conduit and their locations. The National Electrical Code lists where certain conduits, wire and boxes may be installed, but most projects have more restrictive requirements.

The Take-off Package

When all the material quantities have been determined and noted on the pricing sheets, the actual pricing can begin. Keep all pricing sheets in one package; staple them together so that none will be misplaced or lost. Number the pages by indicating the page number and the total number of pages in the package, i.e. Page 1 of 12, Page 2 of 12, and so on. This will show the total count of all the pricing sheets at a glance.

If your take-off package is very large and has dozens of pages, make up an index sheet. This provides quick reference and easy access. An index can also save time when adjustments are made to the material prices. Some suppliers offer additional discounts on large material items. So quick reference to major material items may be useful.

Occasionally changes are made after the take-off has been assembled. The owner may issue a late addendum, adding or deleting certain items. A quick reference will be a great help in making these changes.

Place a cover sheet or end sheet at the front of the take-off package. This sheet is used to list all material prices and the labor hours from the pricing sheets. At this point, labor is still listed in manhours and can be totaled on the cover sheet for conversion to dollars. The design of the cover sheet varies from contractor to contractor. The cover sheet used in this book is suitable for most take-offs.

Chapter 9

Pricing Sheets, Supplier Role

If you have followed me this far, you're now ready to begin compiling material prices. There are many sources for material prices. Some companies publish electrical material prices for electrical contractors and suppliers. In many areas the same service is used by both the contractor and the supplier.

There are usually two pricing books: retail and wholesale. Use the wholesale book for pricing take-offs.

The Price Book
The price service's book probably has thousands of pages of material prices. Index tabs separate the various material groups. New price sheets are provided weekly to keep the service current. Insert them into the book, replacing the old sheets. This can be done by anyone in the office and should always be done as soon as the sheets are received. The sample copies of the service company's wholesale price sheets in this book show how materials are priced.

The service gives prices for most commonly stocked materials. The materials are arranged by the type of material, and in many cases by the manufacturer of the material. Still, many materials won't be listed in the book. The best

way to get those prices is through your supplier. Always allow enough time for the supplier to check with the manufacturers for less common items.

A good electrical pricing service will list the following materials: conduit, elbows, couplings, nipples, locknuts, bushings, straps, insulated wire, bare wire, cable, cords, pigtails, outlet boxes, motor control devices, safety switches, wireway, circuit breakers, condulets, gaskets, junction boxes, lugs, wirenuts, insulated connectors, cable ties, anchors, screws, receptacles, switches, plates, drop cords, construction cords, door chimes, bells, buzzers, burglar alarms. clocks. flashers. transformers. insulators, wireholders, racks, brackets, tapes, splicing materials, fuses, fusetrons, fuseholders, fuse pullers, fans, thermostats, ventilators, incandescent lighting, flood lights, batteries, flasher units, lanterns, lamps, time switches, timers, lighting control, dimmers and photoelectric cells.

The price sheets at the end of this chapter are from a pricing service book by *Trade Service Publications, Inc.*, 10996 Torreyana Road, San Diego, California 92121 and should be used for pricing materials needed on the job in drawings E1 and E2. (See Chapter Seven, Figure 7-1.)

The Supplier

Most electrical material suppliers furnish special price listings to their best customers. The listings give prices for the items they stock. The prices are changed periodically. You should request new price listings from the suppliers at least every six months. Some suppliers offer even better prices on larger orders shipped directly from the manufacturer to the job site.

Usually the price listings furnished by the supplier are for *will-call* material. For the larger direct-shipped orders, the supplier will ask the manufacturer's agent for a special price. Be sure to let your supplier know when you need direct-shipped material pricing.

Sometimes price differences can be traced to the quality of the merchandise. Check the material to be sure that the quality is right for the job. Sometimes bargain items require too much labor to install, or too much effort to get approved.

Additional Services

Suppliers generally offer many additional services to the estimator and the contractor. They will work closely with the estimator to find prices for materials not listed in the trade service books or in their will-call price listings. They'll provide catalogs, special catalog sheets, shop drawings, samples, alternatives, and other valuable assistance.

A good supplier will call manufacturers for sales information or for certain engineering services. They'll process your submittal data quickly so you can get it to the design team for approval. They will follow the material or equipment orders from the manufacturing process to the delivery stage. The supplier's salesmen are well trained to ensure good and prompt service.

But to get this kind of assistance from a supplier, you must be fair in your dealings with him. Don't play one supplier against another. And get the respect of the supplier's salesmen. The salesman may be your only link to a supplier.

Most manufacturers only sell products through the supplier or through an agent working between the manufacturer and the supplier. Most will furnish information only to the supplier on request. A good material supplier will help make your job as an estimator much easier. Find a good supplier and deal fairly with him.

Elec. Page 100-45 A
Jul 1 1981
Cancels Jun 24 1981

PVC PLASTIC CONDUIT AND FITTINGS - PER 100 FEET

DCI EDP NO.	Size	Lbs. Per 100 Feet	Feet Per Bundle	Per 100 Ft.
HEAVY WALL RIGID - SCHEDULE 40				
98-0060-06001	1/2"	17	100	$ 16.39
06002	3/4	23	100	22.27
06003	1	34	100	32.13
06004	1 1/4	46	50	43.50
98-0060-06005	1 1/2	55	50	52.45
06006	2	76	10	69.40
06007	2 1/2	121	10	111.17
06008	3	158	10	144.48
98-0060-06009	3 1/2	190	10	172.88
06010	4	225	10	204.35
06012	5	304	10	290.33
06014	6	394	10	372.84
LIGHT WALL - SCHEDULE A				
98-0060-06021	1/2"	11	100	11.57
06022	3/4	14	100	13.58
06023	1	18	100	17.66
06024	1 1/4	26	50	26.13
98-0060-06025	1 1/2	34	50	33.15
06026	2	53	10	51.06
06028	3	98	10	94.29
06030	4	151	10	150.30

DCI EDP NO.	Size	Description	Std. Pkg.	Bkn. Pkg.	S. P. + Over
ELBOWS - EACH					
98-0060-06041	1/2"	90° Elbow	50	.58	$ .50
06042	3/4	Ditto	25	.64	.56
06043	1	"	25	1.00	.87
06044	1 1/4	"	20	1.39	1.21
98-0060-06045	1 1/2	"	25	1.91	1.66
06046	2	"	15	2.78	2.42
06047	2 1/2	"	5	5.04	4.38
06048	3	"	5	8.82	7.67
98-0060-06049	3 1/2	"	5	12.20	10.61
06050	4	"	5	15.28	13.29
06052	5	"	1	-	23.39
06054	6	"	1	-	39.60
98-0060-06061	1/2	45° Elbows	50	.54	.47
06062	3/4	Ditto	25	.59	.51
06063	1	"	20	.93	.81
06064	1 1/4	"	10	1.30	1.13
98-0060-06065	1 1/2	"	20	1.78	1.55
06066	2	"	10	2.62	2.28
06067	2 1/2	"	5	4.99	4.34
06068	3	"	5	8.49	7.38
98-0060-06069	3 1/2	"	5	9.60	8.35
06070	4	"	5	14.16	12.31
06072	5	"	1	-	19.75
06074	6	"	1	-	29.50
98-0060-06081	1/2	30° Elbow	50	.54	.47
06082	3/4	Ditto	25	.59	.51
06083	1	"	25	.93	.81
06084	1 1/4	"	20	1.29	1.13
98-0060-06085	1 1/2	"	25	1.78	1.55
06086	2	"	20	2.62	2.28
06087	2 1/2	"	5	4.99	4.34
06088	3	"	5	8.49	7.38
98-0060-06089	3 1/2	"	5	9.60	8.35
06090	4	"	5	14.16	12.31
06092	5	"	1	-	19.75
06094	6	"	1	-	23.03

Prices Are Based On Carlon

CASH DISCOUNT 2 %

NOTE: These are samples only. Current prices are available from Trade Service Publications, Inc.
Courtesy: *Trade Service Publications, Inc.*

Sample Pricing Sheets
Figure 9-1

PVC PLASTIC CONDUIT FITTINGS - EACH

Elec. Page 100-46 A
Jul 1 1981
Cancels Jun 24 1981

DCI EDP NO.	Size	Std. Pkg.	Bkn. Pkg.	S. P. + Over
	COUPLINGS			↟
98-0060-06101	1/2"	150	$ .18	$ .16
06102	3/4	100	.23	.20
06103	1	50	.36	.31
06104	1 1/4	30	.47	.41
98-0060-06105	1 1/2	25	.61	.53
06106	2	30	.87	.76
06107	2 1/2	20	1.52	1.32
06108	3	25	2.40	2.09
98-0060-06109	3 1/2	20	2.73	2.37
06110	4	15	3.77	3.28
06112	5	8	9.25	8.04
06114	6	5	12.08	10.50
	EXPANSION COUPLINGS			
	6" Maximum Expansion			
98-0060-06121	1/2"	50	4.28	3.72
06122	3/4	50	4.52	3.93
06123	1	45	4.91	4.27
06124	1 1/4	30	6.95	6.04
98-0060-06125	1 1/2	25	8.50	7.39
06126	2	15	9.89	8.59
06127	2 1/2	10	13.65	11.87
06128	3	10	17.87	15.54
98-0060-06129	3 1/2	5	20.32	17.67
06130	4	5	25.60	22.26
06132	5	3	38.62	33.58
06134	6	2	45.49	39.56
	2" Maximum Expansion			↟
98-0060-06271	1/2"	40	3.94	3.25
06272	3/4	40	3.84	3.34
06273	1	25	4.80	4.17
06274	1 1/4	15	5.98	5.20
98-0060-06275	1 1/2	10	7.79	6.77
06276	2	6	9.63	8.37
	FLEXIBLE ADAPTER COUPLINGS			
98-0060-06238	Small End 3.5 to 4.3" O. D.			
	Large End 4 to 5" O. D.	1	-	11.77 ↟
	REDUCERS			↟
98-0060-06221	3/4 x 1/2"	100	.39	.34
06222	1 x 1/2	100	.47	.41
06223	1 x 3/4	100	.47	.41
06224	1 1/4 x 3/4	50	.70	.61
98-0060-06225	1 1/4 x 1	50	.70	.61
06226	1 1/2 x 1	40	.74	.64
06227	1 1/2 x 1 1/4	40	.74	.64
06228	2 x 1 1/4	25	.87	.76
98-0060-06229	2 x 1 1/2	25	.87	.76
06230	2 1/2 x 2	15	2.83	2.46
06231	3 x 2	25	3.12	2.71
06233	4 x 3	25	4.57	3.97
	FEMALE ADAPTERS			↟
98-0060-06241	1/2"	150	.24	.20
06242	3/4	100	.38	.33
06243	1	50	.52	.45
06244	1 1/4	30	.64	.56
98-0060-06245	1 1/2	25	.70	.61
06246	2	30	1.00	.87
06247	2 1/2	20	1.89	1.64
06248	3	25	2.78	2.42
98-0060-06249	3 1/2	20	3.70	3.22
06250	4	15	4.13	3.59
06252	5	8	10.47	9.10
06254	6	6	12.02	10.45

Prices Are Based On Carlon
CASH DISCOUNT 2%

NOTE: These are samples only. Current prices are available from Trade Service Publications, Inc.
Courtesy: *Trade Service Publications, Inc.*

Elec. Page 114-1 A
Jul 7 1982
Cancels Jun 23 1982

NON-METALLIC SHEATHED CABLE - PER 1000 FEET

600 VOLT

DCI EDP No.	Size	Approx. Net Wt. Per M Ft.	Feet Per Coil	Col. A	Col. B	Col. C
		TYPE NM PLASTIC JACKET				
		Without Ground Wire				†
98-0100-26000	14/2	63 Lbs.	250	$ 122.24	$ 98.58	$ 91.70
26005	12/2	78	250	169.37	136.59	127.06
26010	10/2	109	250	274.82	221.63	206.17
26015	8/2	185	125	601.52	485.09	451.25
26020	6/2	286	125	1014.64	818.26	761.17
98-0100-26050	14/3	89	250	221.40	178.55	166.09
26055	12/3	123	250	308.72	248.97	231.60
26060	10/3	163	250	482.77	389.33	362.17
26065	8/3	271	125	929.05	749.23	696.96
26070	6/3	415	125	1379.59	1112.57	1034.95
26075	4/3	599	125	2048.77	1652.23	1536.96
		With Ground Wire				†
98-0100-26100	14/2	–	250	127.71	103.00	95.81
26105	12/2	–	250	179.44	144.71	134.61
26110	10/2	–	250	318.37	256.75	238.84
26115	8/2	–	125	743.28	599.42	557.60
26120	6/2	–	–	1073.12	865.42	805.04
98-0100-26150	14/3	–	250	242.66	195.69	182.04
26155	12/3	–	250	335.28	270.38	251.52
26160	10/3	–	250	525.44	423.74	394.18
26165	8/3	–	125	1105.59	891.61	829.40
26170	6/3	–	125	1532.32	1235.74	1149.53
98-0100-26175	4/3	–	125	2300.92	1855.58	1726.12
26180	2/3	–	–	3330.35	2685.77	2498.39
26185	14/4	–	–	384.49	310.07	288.44
26188	12/4	–	–	581.67	469.09	436.36
26191	10/4	–	–	768.74	619.95	576.70

NOTE: These are samples only. Current prices are available from Trade Service Publications, Inc.
Courtesy: *Trade Service Publications, Inc.*

Elec. Page 120-103 A
Aug 1 1982
Cancels Jan 20 1982

Mfr's. Zone Western

ALLIED FIBER GLASS OUTLET BOXES + COVERS - PER 100

DCI EDP No.	Cat. No.	Description	Std Pk	Broken Package	Std Pk + Over
		4" ROUND OUTLET BOXES			
78-1193-10584	9364HL	1 5/8" Deep With Bracket	50	ℊ 90.95	ℊ 75.80 ↑
74055	9364HLC1	Ditto W/Brkt + Wire Clamp	50	100.70	83.90 ↑
10585	9364HLC2	" W/Brkt + Wire Clamps	50	110.40	92.00 ↑
10604	9364HLG	" W/Brkt + Grd Strap	50	101.95	84.95 ↑
74065	9364HLGC1	" W/Brkt, Strap + Clamp	50	111.65	93.05 ↑
78-1193-10605	9364HLGC2	" W/Brkt, Strap + Clamps	50	121.45	101.20 ↑
10588	9364Z	" W/Offset Bracket	50	90.95	75.80 ↑
74095	9364ZC1	" W/Brkt + Wire Clamp	50	100.70	83.90 ↑
10589	9364ZC2	" W/Brkt + Wire Clamps	50	110.40	92.00 ↑
10608	9364ZG	" W/Brkt + Grd Strap	50	101.95	84.95 ↑
78-1193-74105	9364ZGC1	" W/Brkt, Strap + Clamp	50	111.65	93.05 ↑
10609	9364ZGC2	" W/Brkt, Strap + Clamps	50	121.45	101.20 ↑
10586	9364Z2	" W/Offset Bracket	50	90.95	75.80 ↑
74115	9364Z2C1	" W/Brkt + Wire Clamp	50	100.70	83.90 ↑
10587	9364Z2C2	" W/Brkt + Wire Clamps	50	110.40	92.00 ↑
78-1193-10606	9364Z2G	" W/Brkt + Grd Strap	50	101.95	84.95 ↑
74125	9364Z2GC1	" W/Brkt, Strap + Clamp	50	111.65	93.05 ↑
10607	9364Z2GC2	" W/Brkt, Strap + Clamps	50	121.45	101.20 ↑
10590	9364-16	" W/Adjustable Bar Hanger	50	136.10	113.40 ↑
74135	9364-16C1	" W/Hanger + Wire Clamp	50	145.80	121.50 ↑
78-1193-10591	9364-16C2	" W/Hanger + Wire Clamps	50	155.60	129.65 ↑
10610	9364-16G	" W/Hanger + Grd Strap	50	147.10	122.60 ↑
74145	9364-16GC1	" W/Hanger + Strap + Clamp	50	156.85	130.70 ↑
10611	9364-16GC2	" W/hanger, Strap + Clamps	50	166.55	138.80 ↑
10592	9364-24	" W/Adjustable Bar Hanger	50	159.00	132.50 ↑
78-1193-74155	9364-24C1	" W/Hanger + Wire Clamp	50	168.70	140.60 ↑
10593	9364-24C2	" W/Hanger + Wire Clamps	50	178.45	148.70 ↑
10612	9364-24G	" W/Hanger + Grd Strap	50	170.00	141.65 ↑
74165	9364-24GC1	" W/Hanger, Strap + Clamp	50	179.75	149.80 ↑
10613	9364-24GC2	" W/Hanger, Strap + Clamps	50	189.50	157.90 ↑
78-1193-10630	9365N	" W/Nails	50	80.15	66.80 ↑
10631	9365NC1	" W/Nails + Wire Clamp	50	85.25	71.05 ⊕
10632	9365NC2	" W/Nails + Wire Clamps	50	90.35	75.30 ↓
10633	9365NG	" W/Nails + Grd Strap	50	91.75	76.45 ↑
10634	9365NGC1	" W/Nails, Strap + Clamp	50	101.90	84.90 ↑
78-1193-10635	9365NGC2	" W/Nails, Strap + Clamps	50	112.15	93.45 ↑
10665	9365-16	" W/Adjustable Bar Hanger	50	148.80	124.00 ↑
10666	9365-16C1	" W/Hanger + Wire Clamp	50	159.05	132.55 ↑
10667	9365-16C2	" W/Hanger + Wire Clamps	50	169.30	141.10 ↑
10668	9365-16G	" W/Ground Strap	50	160.40	133.65 ↑
78-1193-10669	9365-16GC1	" W/Hanger, Strap + Clamp	50	170.65	142.20 ↑
10670	9365-16GC2	" W/Hanger, Strap + Clamps	50	180.85	150.70 ↑
10671	9365-24	" W/Adjustable Bar Hanger	50	163.80	135.50 ↑
10672	9365-24C1	" W/Hanger + Wire Clamp	50	173.95	144.95 ↑
10673	9365-24C2	" W/Hanger + Wire Clamps	50	184.20	153.50 ↑
78-1193-10674	9365-24G	" W/Hanger + Grd Strap	50	175.25	146.05 ↑
10675	9365-24GC1	" W/Hanger, Strap + Clamp	50	185.50	154.60 ↑
10676	9365-24GC2	" W/Hanger, Strap + Clamps	50	195.80	163.15 ↑
	ROUND BOX COVERS				↑
78-1193-10130	9315	4 1/2" Blank	100	42.10	35.10
10131	9316	4 1/2 Duplex Receptacle	100	49.40	41.15
10132	9317	4 1/2 Toggle Switch	100	49.40	41.15
10366	9348	4" Plaster Ring With 7/16" Offset	100	44.15	36.80

⊕ - Item No Longer Listed In Mfr's Price Sheet

NOTE: These are samples only. Current prices are available from Trade Service Publications, Inc.
Courtesy: *Trade Service Publications, Inc.*

Elec. Page 120-105 B
Aug 1 1982
Cancels Jan 20 1982

Mfr's. Zone Western

ALLIED FIBER GLASS SWITCH BOXES + COVERS - PER 100

DCI EDP No.	Cat. No.	Description	Std Pk	Broken Package	Std Pk + Over
2 GANG SWITCH BOXES					
78-1193-10045	9308E	1 1/2" Deep With Ears	50	$ 80.70	$ 67.25 ↑
10065	9312	2 1/2 Deep	50	87.25	72.70 ↑
28055	9312C2	Ditto With Wire Clamps	50	106.75	88.95 ↑
10066	9312C4	" With Wire Clamps	50	126.20	105.15 ↑
10064	9312E	" With Ears	50	84.00	70.00 ↑
78-1193-28075	9312EC2	" With Ears + Wire Clamps	50	103.50	86.25 ↑
10068	9312EC4	" With Ears + Wire Clamps	50	123.30	102.75 ↑
29100	9312EW	" With Ears + Brkt	50	106.70	88.90 ↑
29110	9312EWC2	" W/Ears, Brkt + Clamps	50	115.75	96.45 ↑
29120	9312EWC4	" W/Ears, Brkt + Clamps	50	135.25	112.70 ↑
78-1193-10067	9312EX	" With Ears	50	84.00	70.00 ↑
10069	9312HF	" With Bracket	50	111.40	92.85 ↑
28095	9312HFC2	" With Brkt + Wire Clamps	50	130.85	109.05 ↑
10070	9312HFC4	" With Brkt + Wire Clamps	50	150.35	125.30 ↑
10062	9312HN	" With Bracket	50	111.40	92.85 ↑
78-1193-28115	9312HNC2	" With Brkt + Wire Clamps	50	130.85	109.05 ↑
10063	9312HNC4	" With Brkt + Wire Clamps	50	150.35	125.30 ↑
10071	9312Z2	" Deep With Offset Brkt	50	111.40	92.85 ↑
28135	9312Z2C2	" With Brkt + Clamps	50	130.85	109.05 ↑
10074	9312Z2C4	" With Brkt + Clamps	50	150.35	125.30 ↑
78-1193-10072	9312-2Z2	" With 2 Offset Brkts	25	135.60	113.00 ↑
28155	9312-2Z2C2	" With 2 Brkts + Clamps	25	155.05	129.20 ↑
10077	9312-2Z2C4	" With 2 Brkts + Clamps	25	174.55	145.45 ↑
10680	9372	2 3/4" Deep Less Ears	50	101.75	84.80 D
10700	9372HF	Ditto With Brkt	50	139.55	116.30 D
78-1193-10697	9372HFB	" With Brkt + Bar Hanger	50	156.50	130.40 D
10681	9372N	" With Nails	50	108.90	90.75 D
10682	9372NMH	" With Nails (Mobile Home)	50	99.05	82.55 D
77310	9372NT	" With Nails + Speed Clip	50	120.25	100.20 D
10685	9372Z2	" With Offset Brkt	50	139.55	116.30 D
78-1193-77065	9372Z2B	" With Brkt + Bar Hanger	50	156.50	130.40 D
10690	9372-2Z2	" With 2 Offset Brkts	25	156.50	130.40 D
3 GANG SWITCH BOXES					↑
78-1193-14175	3300N	3" Deep With Nails	20	167.90	139.90
14185	3300NC2	Ditto With Nails + 2 Clamps	20	187.40	156.15
14190	3300NC3	" With Nails + 3 Clamps	20	197.10	164.25
14195	3300NC4	" With Nails + 4 Clamps	20	206.80	172.35
14205	3300NC6	" With Nails + 6 Clamps	20	226.25	188.55
78-1193-14210	3300NB	" With Nails + Hanger	20	184.90	154.10
14220	3300NBC2	" With Nails + 2 Clamps	20	204.40	170.35
14225	3300NBC3	" With Nails + 3 Clamps	20	214.15	178.45
14230	3300NBC4	" With Nails + 4 Clamps	20	223.85	186.55
14240	3300NBC6	" With Nails + 6 Clamps	20	243.30	202.75
78-1193-14560	3300Z2	" With Offset Bracket	25	187.50	156.25
14570	3300Z2C2	" With Brkt + 2 Clamps	25	206.95	172.45
14575	3300Z2C3	" With Brkt + 3 Clamps	25	216.65	180.55
14580	3300Z2C4	" With Brkt + 4 Clamps	25	226.45	188.70
14590	3300Z2C6	" With brkt + 6 Clamps	25	245.90	204.90
78-1193-14595	3300Z2B	" With Brkt + Hanger	20	204.55	170.45
14605	3300Z2BC2	" W/Brkt, Hanger + 2 Clamps	20	224.00	186.65
14610	3300Z2BC3	" W/Brkt, Hanger + 3 Clamps	20	233.75	194.80
14615	3300Z2BC4	" W/Brkt, Hanger + 4 Clamps	20	243.50	202.90
14625	3300Z2BC6	" W/Brkt, Hanger + 6 Clamps	20	262.90	219.10
78-1193-14700	3300 2Z2	" With Offset Brackets	25	204.55	170.45
14710	33002Z2C2	" With Brkts + 2 Clamps	25	224.00	186.65
14715	33002Z2C3	" With Brkts + 3 Clamps	25	233.75	194.80
14720	33002Z2C4	" With Brkts + 4 Clamps	25	243.50	202.90
14730	33002Z2C6	" With brkts 7 6 Clamps	25	262.90	219.10
78-1193-10087	9313E	2 1/2" Deep With Ears	50	150.60	125.50
30045	9313EC2	Ditto With Ears + Wire Clamps	50	170.05	141.70
30055	9313EC4	" With Ears + Wire Clamps	50	189.55	157.95
10088	9313EC6	" With Ears + Wire Clamps	50	209.00	174.15

D - Discontinued Item

NOTE: These are samples only. Current prices are available from Trade Service Publications, Inc.
Courtesy: *Trade Service Publications, Inc.*

Mfr's. Zone 3

Elec. Page 120-151 A
Nov 9 1981
Cancels May 20 1981

APPLETON OUTLET BOXES AND COVERS - PER 100

DCI EDP NO.	Cat. No.	Universal Key No.	Description	Std. Pkg.	Bkn. Pkg.	S. P. + Over
	3" OCTAGON OUTLET BOXES					
78-1381-75435	30 1/2	24151 1/2	1 1/2" Deep - 1/2" KO	50	$116.25	$ 93.00 †
	3" OCTAGON EXTENSION RINGS					
78-1381-75440	30E 1/2	24151 1/2	1 1/2" Deep - 1/2" KO	25	139.81	111.85 †
	3" ROUND BOX COVERS					†
78-1381-77455	8301A	24C1	Flat Blank	50	41.88	33.50
77465	8320	24C6	Flat - 1/2" KO	50	42.81	34.25
	4" OCTAGON OUTLET BOXES					†
78-1381-75000	40 1/2	54151 1/2	1 1/2" Deep-1/2" KO	50	103.06	82.45
75010	40 3/4	54151 3/4	1 1/2" Deep-3/4 KO	50	105.19	84.15
75475	40SPL	54151-SPL	1 1/2 Deep-1/2 + 3/4 KO	50	105.19	84.15
75015	40D 1/2	54171 1/2	2 1/8" Deep-1/2" KO	25	130.13	104.10
78-1381-75020	40D 3/4	54171 3/4	2 1/8 Deep-3/4 KO	25	139.56	111.65
75025	40D-SPL	54171-SPL	2 1/8 Deep-1/2 + 3/4 KO	25	139.56	111.65
75055	40JB 1/2	54151-J 1/2	1 1/2 Deep w/Bracket	25	135.19	108.15
75480	40VB 1/2	54151-F 1/2	1 1/2 Deep w/Bracket	50	129.81	103.85
	4" OCTAGON EXTENSION RINGS					†
78-1381-75030	40E 1/2	55151 1/2	1 1/2" Deep-1/2" KO	50	123.94	99.15
75035	40E 3/4	55151 3/4	1 1/2 Deep-3/4 KO	50	125.19	100.15
75050	40E-SPL	55151	1 1/2 Deep-1/2 + 3/4 KO	50	128.13	102.50
	4" ROUND BOX COVERS					†
78-1381-77525	8403	54C1	Flat Blank	50	36.88	29.50
77530	8409	54C3	Plaster Ring Raised-5/8"	50	71.75	57.40
77535	8409A	54C3 1/2	Plaster Ring Raised-1/2"	50	68.31	54.65
77555	8413	54C6	Flat - 1/2" KO	50	36.44	29.15
77565	8419LR	54S	Flat For Single Recept.	50	77.50	62.00
77570	8420LR	54D	Flat For Duplex Recept.	50	77.50	62.00
	4" SQUARE OUTLET BOXES					†
78-1381-75065	4S 1/2	52151 1/2	1 1/2" Deep-1/2" KO	50	123.00	98.40
75070	4S 3/4	52151 3/4	1 1/2 Deep-3/4 KO	50	126.81	101.45
75335	4S-SPL	52151-SPL	1 1/2 Deep-1/2 + 3/4 KO	50	126.81	101.45
17539	4SP	-	1 1/2 Deep-Plenum	50	261.00	208.80
75100	4SD 1/2	52171 1/2	2 1/8 Deep-1/2" KO	25	203.88	163.10
78-1381-75105	4SD 3/4	52171 3/4	2 1/8 Deep-3/4 KO	25	209.81	167.85
75095	4SD1	52171-1	2 1/8 Deep-1	25	209.81	167.85
75110	4SD-SPL	52171-SPL	2 1/8 Deep-1/2 + 3/4 KO	25	209.81	167.85
17541	4SP-D	-	2 1/8 Deep-Plenum	25	421.31	337.05
17542	4SP-JD	-	2 1/8 Deep-Plenum	25	528.13	422.50
75230	4SL 1/2	52141 1/2	1 1/4 Deep-1/2" KO	50	144.19	115.35
	4" SQUARE BOXES WITH BRACKETS					†
78-1381-19069	4SAB 1/2	-	1 1/2" Deep-1/2" KO	25	158.63	126.90
28944	4SAB-SPL	-	1 1/2 Deep-1/2 + 3/4" KO	25	164.06	131.25
75080	4SB 1/2	52151B 1/2	1 1/2 Deep-1/2" KO	25	164.19	131.35
28946	4SDVB1/2PL	-	2 1/8 Deep-1/2 KO	25	241.94	193.55
28947	4SDVB-SPL-PL	-	2 1/8 Deep-1/2 + 3/4 KO	25	247.75	198.20
78-1381-75250	4SOB 1/2	52151-O 1/2	1 1/2 Deep-1/2" KO	25	161.44	129.15
25158	4SRAB	-	1 1/2 Deep w/Romex Clamps	25	196.44	157.15
75260	4SRB	52151-BN	1 1/2 Deep Ditto	25	214.81	171.85
17544	4SRDVB	-	2 1/8 Deep "	25	276.06	220.85
17546	4SROB 1/2	-	1 1/2 Deep "	25	196.56	157.25
78-1381-75265	4SRVB	52151-FN	1 1/2 Deep "	25	196.44	157.15
75345	4SVB 1/2	52151-F 1/2	1 1/2 Deep-1/2" KO)	25	158.63	126.90
75340	4SVB 1/2-PL	52151F 1/2	1 1/2 Deep-1/2 KO	25	158.63	126.90
75355	4SVB-SPL-PL	52151F-SPL	1 1/2 Deep-1/2 + 3/4 KO	25	164.06	131.25
75270	4SRVB-PL	52151F-N	1 1/2 Deep-1/2 KO	25	196.44	157.15
78-1381-28948	4SXAB	-	1 1/2 Deep w/Bx Clamps	25	225.88	180.70
75360	4SXVB	52151-FX	1 1/2 Deep w/Bx Clamps	25	225.88	180.70
75350	4SVB-SPL	52151-F-SPL	1 1/2 Deep w/Bx Clamps	25	164.06	131.25

NOTE: These are samples only. Current prices are available from Trade Service Publications, Inc.
Courtesy: *Trade Service Publications, Inc.*

Elec. Page 120-151 B
Nov 9 1981
Cancels May 20 1981

Mfr's. Zone 3

APPELTON OUTLET BOXES AND COVERS - PER 100

DCI EDP NO.	Cat. No.	Universal Key No.	Description	Std. Pkg.	Bkn. Pkg.	S. P. + Over
4" SQUARE BOXES - WITH CLAMPS - WITHOUT BRACKETS						†
78-1381-75255	4SR	52151N	1 1/2" Deep w/Bx Clamps	25	$178.81	$143.05
17543	4SRD	-	2 1/8 Deep w/Romex Clamps	25	258.38	206.70
4" SQUARE BOX EXTENSIONS						†
78-1381-75125	4SES	-	1 1/2 Deep - 1/2 + 3/4" KO	50	163.06	130.45
75115	4SE1/2	53151 1/2	1 1/2 Deep - 1/2 KO	50	159.81	127.85
75120	4SE3/4	53151	1 1/2 Deep - 3/4 KO	50	163.06	130.45
4" SQUARE BOX COVERS						†
78-1381-77620	8461	52C3	Plaster Ring Raised - 5/8"	50	72.69	58.15
77625	8461A	52C3-1/2	Plaster Ring Raised - 1/2	50	69.94	55.95
77630	8461B	52C3-3/4	Plaster Ring Raised - 3/4	25	91.25	73.00
77635	8461C	52C3-1	Plaster Ring Raised - 1	50	243.94	195.15
77640	8461D	52C3-1 1/4	Plaster Ring Raised - 1 1/4	25	243.94	195.15
78-1381-77645	8465	52C1	Flat Blank	50	48.81	39.05
17549	8465P	-	Flat Blank w/Gasket	50	126.25	101.00
77650	8466	52C16	1 1/4" Single Switch Ring	25	105.19	84.15
77655	8466A	52C15	1 Single Switch Ring	25	98.69	78.95
77660	8468	52C14	3/4 Single Switch Ring	25	75.25	60.20
78-1381-77665	8468A	52C13	1/2 Single Switch Ring	25	63.63	50.90
77670	8468B	52C62	1/4 Single Switch Ring	25	63.63	50.90
77675	8468C	52C14-5/8	5/8 Single Switch Ring	50	74.75	59.80
77680	8468F	52C0	Flat Single Switch Ring	50	66.94	53.55
77685	8468WB	52C36	1/4" Single Switch Ring	50	63.63	50.90
78-1381-77695	8469A	52C19	1 Two G Switch Ring	25	156.06	124.85
77700	8470	52C18	3/4 Two G Switch Ring	50	112.19	89.75
77705	8470A	52C17	1/2 Two G Switch Ring	50	112.19	89.75
77710	8470B	52C20	1/4 Two G Switch Ring	50	112.19	89.75
77715	8470C	5217	5/8 Two G Switch Ring	50	115.44	92.35
78-1381-77720	8470F	52C00	Flat Two G Switch Ring	50	87.94	70.35
77725	8470WB	52C37	1/4" Two G Switch Ring	50	86.56	69.25
77730	8474	52C6	Flat - 1/2" KO	50	47.38	37.90
TILE COVERS FOR 4" SQUARE BOXES						†
78-1381-76825	846-50	52C49-1/2	1/2" Single Switch Ring	25	158.56	126.85
76830	846-75	52C49-3/4	3/4 Ditto	25	163.63	130.90
76835	846-100	52C49-1	1 "	25	168.50	134.80
76840	846-125	52C49-1 1/4	1 1/4 "	25	173.50	138.80
76845	846-150	52C50-1 1/2	1 1/2 "	25	177.81	142.25
78-1381-76850	846-200	52C51-2	2 "	25	183.38	146.70
76855	847-50	52C52-1/2	1/2 Two G Switch Ring	25	203.31	162.65
76860	847-75	52C52-3/4	3/4 Ditto	25	203.31	162.65
76865	847-100	52C52-1	1 "	25	203.31	162.65
76870	847-125	52C52-1 1/4	1 1/4 "	25	203.31	162.65
78-1381-76875	847-150	52C53-1 1/2	1 1/2 "	25	211.44	169.15
76880	847-200	52C54-2	2 "	25	211.44	169.15
SURFACE COVERS FOR 4" SQUARE BOXES						†
78-1381- -	8360	-	1/2" Deep Blank	25	106.50	85.20 A
77470	8361	53-1T	1/2 " 1 Toggle Switch	25	106.50	85.20
77475	8363	52-1S	1/2 " 1 Flush Recept.	25	106.50	85.20
-	8364	-	1/2" " 2 Flush Recept.	25	106.50	85.20 A
77480	8365	52-1D	1/2 " 1 Duplex Recept.	25	106.50	85.20
78-1381 77485	8367	52-2T	1/2 " 2 Toggle Switches	25	106.50	85.20
-	8368	-	1/2 " 1 Tgl Sw + Fl Rep	25	106.50	85.20 A
77495	8371	52-2D	1/2 " 2 Duplex Recepts.	25	106.50	85.20
77500	8375	52-TD	1/2 " 1 Toggle + Duplex	25	106.50	85.20
77505	8377	-	1/2 " 1 30-50A Recept.	25	106.50	85.20
78-1381- -	8378	-	1/2" " 1 Single Recept.	25	106.50	85.20 A
77510	8379	-	1/2 " 1 40W Twist Lock	25	106.50	85.20

A - Added Listing

Elec. Page 168-1 B
Oct 19 1981
Cancels Mar 2 1981

LEVITON WIRING DEVICES - PER 100

DCI EDP NO.	Cat. No.	Car-ton	Std. Pkg.	List	Bkn. Ctn.	Ctn. + Over	S. P. + Over	
78-4146-10832	223-ISP	10	200	$105.50	$94.95	$84.40	$73.85	↑
10822	223-SP	10	200	102.50	92.25	82.00	71.75	↑
10837	223-WSP	10	200	110.00	99.00	88.00	77.00	↑
10900	224	10	100	105.00	94.50	84.00	73.50	↑D
10910	224-I	10	100	110.50	99.45	88.40	77.35	↑D
78-4146-10920	228	10	100	148.00	133.20	118.40	103.60	↑D
10930	229	10	100	161.50	145.35	129.20	113.05	↑
10940	233	1	50	469.50	–	375.60	328.65	↑D
10950	237	1	50	649.50	–	519.60	454.65	↑D
10960	238	10	100	118.50	106.65	94.80	82.95	↑
78-4146-10970	238-I	10	100	122.00	109.80	97.60	85.40	↑
11000	242	10	200	137.50	123.75	110.00	96.25	↑
11010	242-I	10	200	139.50	125.55	111.60	97.65	↑
11140	267	25	100	82.00	73.80	65.60	57.40	↑
11150	267-2	25	100	72.50	65.25	58.00	50.75	D
78-4146-11160	274	10	100	76.00	68.40	60.80	53.20	↑
11165	274B	10	100	76.00	68.40	60.80	53.20	↑
11170	274K	10	100	87.50	78.75	70.00	61.25	↑
11183	274YL	10	100	76.00	68.40	60.80	53.20	↑
11185	275	1	10	1129.50	–	903.60	790.65	↑
78-4146-11190	276	1	50	1116.50	–	893.20	781.55	↑
11200	276-2	1	50	1116.50	–	893.20	781.55	↑D
11210	277	1	50	1116.50	–	893.20	781.55	↑
11215	277GY	1	50	1116.50	–	893.20	781.55	↑D
11220	277-2	1	50	1116.50	–	893.20	781.55	↑D
78-4146-11225	277-2GY	1	50	1116.50	–	893.20	781.55	↑D
11227	277-7	1	50	1116.50	–	893.20	781.55	↑
11230	278	1	10	1366.50	–	1093.20	956.55	↑
11235	278GY	1	10	1366.50	–	1093.20	956.55	↑
11240	279	1	10	1366.50	–	1093.20	956.55	↑
78-4146-11243	279GY	1	10	1366.50	–	1093.20	956.55	↑
11250	283-3	1	10	695.00	–	556.00	486.50	↑
11260	283-4	1	10	807.50	–	646.00	565.25	↑
11270	283-5	1	10	909.50	–	727.60	636.65	↑
11280	283-6	1	10	1021.50	–	817.20	715.05	↑
78-4146-11290	284-3	1	10	855.00	–	684.00	598.50	↑
11300	284-4	1	10	1014.00	–	811.20	709.80	↑
11310	284-5	1	10	1170.00	–	936.00	819.00	↑
11320	284-6	1	10	1301.50	–	1041.20	911.05	↑
11340	285-3	1	10	640.00	–	512.00	448.00	↑
78-4146-11350	285-4	1	10	720.50	–	576.40	504.35	↑
11360	285-5	1	10	830.00	–	664.00	581.00	↑
11370	285-6	1	10	877.50	–	702.00	614.25	↑
11390	287	1	10	712.50	–	570.00	498.75	↑
11400	302	10	250	115.50	103.95	92.40	80.85	↑
78-4146-11410	302-I	10	250	119.50	107.55	95.60	83.65	↑
11420	303	10	250	115.50	103.95	92.40	80.85	↑
11430	303-I	10	250	119.50	107.55	95.60	83.65	↑
11460	306	10	100	98.50	88.65	78.80	68.95	↑
11470	306W	10	100	101.50	91.35	81.20	71.05	↑
78-4146-11510	314	10	100	153.00	137.70	122.40	107.10	↑
11520	314W	10	100	170.00	153.00	136.00	119.00	↑

D - To be discontinued when present stock is exhausted.

NOTE: These are samples only. Current prices are available from Trade Service Publications, Inc.
Courtesy: *Trade Service Publications, Inc.*

Elec. Page 168-3 B
Dec 23 1981
Cancels Oct 19 1981

LEVITON WIRING DEVICES -- PER 100

DCI EDP NO.	Cat. No.	Carton	Std. Pkg.	List	Bkn. Cart.	Cart. + Over	S. P. + Over	
78-4146-13380	530K	10	100	$ 82.00	$ 73.80	$ 65.60	$ 57.40	
13390	530-IK	10	100	82.00	73.80	65.60	57.40	
13550	554	Bulk	200	622.00	-	497.60	435.40	
13560	555	Bulk	200	622.00	559.80	497.60	435.40	
13570	565	10	100	179.00	161.10	143.20	125.30	
78-4146-13580	566	10	100	183.50	165.15	146.80	128.45	
13610	572T	10	100	118.00	106.20	94.40	82.60	
13612	572T30	10	100	118.00	106.20	94.40	82.60	
13640	573	10	100	166.00	149.40	132.80	116.20	
13650	573-1	10	100	193.00	173.70	154.40	135.10	D
78-4146-13675	573-17	10	100	214.50	193.05	171.60	150.15	D
13680	574T	10	100	117.50	105.75	94.00	82.25	
13700	575	10	100	139.00	125.10	111.20	97.30	
13750	577T	10	100	134.50	121.05	107.60	94.15	
13752	577T50	10	100	134.50	121.05	107.60	94.15	
78-4146-13780	579	10	100	144.50	130.05	115.60	101.15	
13790	579-1	10	100	150.50	135.45	120.40	105.35	D
13810	581	10	100	206.00	185.40	164.80	144.20	
13813	581-27	10	100	206.00	185.40	164.80	144.20	x
13850	585	10	100	194.00	174.60	155.20	135.80	
78-4146-13870	587	10	100	247.50	222.75	198.00	173.25	
13890	591	10	100	206.00	185.40	164.80	144.20	D
13900	601	1	24	616.00	-	492.80	431.20	
13910	602	1	24	554.50	-	443.60	388.15	
13920	603	1	24	757.00	-	605.60	529.90	
78-4146-13980	612	10	100	139.50	125.55	111.60	97.65	
14020	617	10	100	359.50	323.55	297.60	251.65	
14030	628	1	50	180.00	-	144.00	126.00	
14040	628-I	1	50	183.00	-	146.40	128.10	
14055	637B	25	250	71.50	64.35	57.20	50.05	D
78-4146-14060	637W	25	250	71.50	64.35	57.20	50.05	D
14070	638	25	250	79.00	71.10	63.20	55.30	
14075	638B	25	250	79.00	71.10	63.20	55.30	
14080	638W	25	250	79.00	71.10	63.20	55.30	
14155	649B	25	250	143.50	129.15	114.80	100.45	
78-4146-14160	649W	25	250	156.00	140.40	124.80	109.20	
14185	654	Bulk	250	Discontinued - See No. 6254				
14230	658	1	50	285.50	-	228.40	199.85	
14240	658BR	1	50	285.50	-	228.40	199.85	
14250	658-I	1	50	289.00	-	231.20	202.30	
78-4146-14255	659	Bulk	250	Discontinued - See No. 6259				
14270	663	10	100	191.50	172.35	153.20	134.05	
14300	679	10	100	136.00	122.40	108.80	95.20	
14330	681	25	250	199.50	179.55	159.60	139.65	D
14340	681W	25	250	205.00	184.50	164.00	143.50	D
78-4146-14400	685	10	100	304.00	273.60	243.20	212.80	D
14413	688	1	50	289.00	-	231.20	202.30	
14416	688-I	1	50	292.50	-	234.00	204.75	
14420	689	10	100	495.50	445.95	396.40	346.85	
14450	690	10	100	509.00	458.10	407.20	356.30	
14480	691	10	100	495.50	445.95	396.40	346.85	

D - To be discontinued when present stock is exhausted.
x - Price Correction

LEVITON WIRING DEVICES -- PER 100

DCI EDP NO.	Cat. No.	Car-ton	Std. Pkg.	List	Bkn. Cart.	Cart. + Over	S. P. + Over	
78-4146-16930	1331-I	10	100	$188.50	$169.65	$150.80	$131.95	
16932	1331-ISP	10	100	188.50	169.65	150.80	131.95	
16922	1331SP	10	100	183.00	164.70	146.40	128.10	
16940	1331-2	10	100	197.50	177.75	158.00	138.25	
16950	1331-2-I	10	100	202.50	182.25	162.00	141.75	
78-4146-17017	1373	Bulk	250	194.00	–	155.20	135.80	
17019	1373W	10	100	195.00	175.50	156.00	136.50	
17025	1374-1	10	100	149.50	134.55	119.60	104.65	
17035	1374-1W	10	100	164.00	147.60	131.20	114.80	
17120	1403	10	250	110.50	99.45	88.40	77.35	
78-4146-17130	1403-I	10	250	113.50	102.15	90.80	79.45	
17140	1406	10	200	164.50	148.05	131.60	115.15	
17150	1406-I	10	200	169.00	152.10	135.20	118.30	
	1419	10	100	234.00	210.60	187.20	163.80	
	1419-I	10	100	234.00	210.60	187.20	163.80	
78-4146-	1419W	10	100	234.00	210.60	187.20	163.80	
17223	1421	10	100	210.50	189.45	168.40	147.35	
17225	1421-I	10	100	214.00	192.60	171.20	149.80	
17227	1421W	10	100	215.00	193.50	172.00	150.50	
17230	1430	10	100	118.50	106.65	94.80	82.95	D
78-4146-17240	1430-I	10	100	122.00	109.80	97.60	85.40	D
17280	1431	10	100	188.50	169.65	150.80	131.95	D
17290	1431-I	10	100	192.00	172.80	153.60	134.40	D
17300	1432	1	50	524.50	–	419.60	367.15	
17310	1433	1	50	628.00	–	502.40	439.60	
78-4146-17332	1440	Bulk	20	115.50	–	92.40	80.85	
17340	1451	10	100	210.50	189.45	168.40	147.35	
17341	1451-CP	10	100	210.00	189.00	168.00	147.00	
17350	1451-I	10	100	214.00	192.60	171.20	149.80	
17351	1451-I-CP	10	100	213.50	192.15	170.80	149.45	
78-4146-17352	1451-ISP	10	100	214.00	192.60	171.20	149.80	
17342	1451SP	10	100	210.50	189.45	168.40	147.35	
17360	1451W	10	100	215.00	193.50	172.00	150.50	
17362	1451WSP	10	100	215.00	193.50	172.00	150.50	
17370	1451-2	10	100	221.00	198.90	176.80	154.70	
78-4146-17380	1451-2-I	10	100	224.50	202.05	179.60	157.15	
17390	1451-2W	10	100	226.00	203.40	180.80	158.20	
17400	1451-4	Bulk	100	221.00	–	176.80	154.70	
17410	1451-4-I	Bulk	100	224.50	–	179.60	157.15	
17415	1451-4W	Bulk	100	226.00	–	180.80	158.20	
78-4146-17420	1453	10	100	337.00	303.30	269.60	235.90	
17421	1453-CP	10	100	336.50	302.85	269.20	235.55	
17435	1453-I	10	100	341.00	306.90	272.80	238.70	
17436	1453-I-CP	10	100	340.50	306.45	272.40	238.35	x
17432	1453-ISP	10	100	341.00	306.90	272.80	238.70	
78-4146-17422	1453SP	10	100	337.00	303.30	269.60	235.90	
17440	1453W	10	100	342.00	307.80	273.60	239.40	
17442	1453WSP	10	100	342.00	307.80	273.60	239.40	
17450	1453-2	10	100	347.50	312.75	278.00	243.25	
17460	1453-2-I	10	100	351.00	315.90	280.80	245.70	
78-4146-17470	1453-2W	10	100	352.00	316.80	281.60	246.40	
17480	1453-4	Bulk	100	347.50	–	278.00	243.25	

D - To Be Discontinued When Present Stock Is Exhausted.
x - Price Correction

NOTE: These are samples only. Current prices are available from Trade Service Publications, Inc.
Courtesy: *Trade Service Publications, Inc.*

Elec. Page 168-17 A
Oct 19 1981
Cancels Mar 2 1981

LEVITON WIRING DEVICES -- PER 100

DCI EDP NO.	Cat. No.	Carton	Std. Pkg.	List	Bkn. Ctn.	Ctn. + Over	S. P. + Over
78-4146-21930	5175	10	30	$ 798.00	$ 718.20	$ 638.40	$ 558.60
21950	5176	10	20	727.00	654.30	581.60	508.90
21980	5177	10	20	1044.50	940.05	835.60	731.15
22000	5178	10	20	598.50	538.65	478.80	418.95
22060	5181	10	20	1194.00	1074.60	955.20	835.80
78-4146-22070	5181G	10	20	1194.00	1074.60	955.20	835.80
22080	5182	10	20	782.50	704.25	626.00	547.75
22090	5182G	10	20	782.50	704.25	626.00	547.75
22130	5185	10	50	1093.50	984.15	874.80	765.45
22150	5190	10	20	1237.50	1113.75	990.00	866.25
78-4146-22250	5195	10	20	1096.50	986.85	877.20	767.55
22270	5198	10	20	843.50	759.15	674.80	590.45
22310	5206	1	20	498.00	-	398.40	348.60
22315	5206GY	1	20	505.00	-	404.00	353.50 D
22320	5207	1	20	498.00	-	398.40	348.60
78-4146-22325	5207GY	1	20	505.00	-	404.00	353.50 D
22330	5211	10	100	296.00	266.40	236.80	207.20
22340	5211-I	10	100	301.00	270.90	240.80	210.70
22342	5211-ISP	10	100	301.00	270.90	240.80	210.70
22332	5211SP	10	100	296.00	266.40	236.80	207.20
78-4146-22350	5212	10	100	303.50	273.15	242.80	212.45
22360	5212-I	10	100	308.00	277.20	246.40	215.60
22362	5212-ISP	10	100	308.00	277.20	246.40	215.60
22352	5212SP	10	100	303.50	273.15	242.80	212.45
22365	5212-2-I	10	100	316.00	284.40	252.80	221.20
78-4146-22370	5213LB	10	100	348.50	313.65	278.80	243.95
22380	5213LB-I	10	100	351.00	315.90	280.80	245.70
22390	5214	10	100	312.50	281.25	250.00	218.75
22400	5214-I	10	100	315.50	283.95	252.40	220.85
22410	5215	10	50	463.50	417.15	370.80	324.45
78-4146-22420	5215-I	10	50	466.50	419.85	373.20	326.55
22430	5216	10	50	477.50	429.75	382.00	334.25
22440	5216-I	10	50	481.50	433.35	385.20	337.05
22450	5217	10	50	509.50	458.55	407.60	356.65
22460	5217-I	10	50	513.00	461.70	410.40	359.10
78-4146-22470	5218	10	50	509.50	458.55	407.60	356.65
22480	5218-I	10	50	513.00	461.70	410.40	359.10
22490	5219	10	100	517.00	465.30	413.60	361.90
22500	5219-I	10	100	520.50	468.45	416.40	364.35
22510	5220	10	100	521.00	468.90	416.80	364.70
78-4146-22520	5220-I	10	100	523.50	471.15	418.80	366.45
22530	5221	10	100	303.50	273.15	242.80	212.45
22540	5221-I	10	100	308.00	277.20	246.40	215.60
22560	5222	10	100	464.00	417.60	371.20	324.80
22570	5222-I	10	100	467.00	420.30	373.60	326.90
78-4146-22572	5222-ISP	10	100	467.00	420.30	373.60	326.90
22562	5222SP	10	100	464.00	417.60	371.20	324.80
22582	5222WSP	10	100	470.00	423.00	376.00	329.00
22620	5224	10	100	617.00	555.30	493.60	431.90

D - To Be Discontinued When Present Stock is Exhausted

NOTE: These are samples only. Current prices are available from Trade Service Publications, Inc.
Courtesy: *Trade Service Publications, Inc.*

Elec. Page 168-17 B
Oct 19 1981
Cancels Mar 2 1981

				LEVITON WIRING DEVICES -- PER 100			†
DCI EDP NO.	Cat. No.	Car-ton	Std. Pkg.	List	Bkn. Ctn.	Ctn. + Over	S. P. + Over
78-4146-22630	5224-I	10	100	$ 619.50	$ 557.55	$ 495.60	$ 433.65
22632	5224-ISP	10	100	619.50	557.55	495.60	433.65
22622	5224-SP	10	100	617.00	555.30	493.60	431.90
22642	5224WSP	10	100	622.00	559.80	497.60	435.40
22643	5224-2	10	100	651.50	586.35	521.20	456.05
78-4146-22645	5224-2-I	10	100	655.50	589.95	524.40	458.85
22648	5224-2W	10	100	663.00	596.70	530.40	464.10
22650	5225	10	100	655.50	589.95	524.40	458.85
22660	5225-I	10	100	660.00	594.00	528.00	462.00
22662	5225-ISP	10	100	660.00	594.00	528.00	462.00
78-4146-22652	5225SP	10	100	655.50	589.95	524.40	458.85
22672	5225WSP	10	100	663.00	596.70	530.40	464.10
22675	5225-49	1	50	1065.00	-	852.00	745.50
22680	5226	10	100	530.00	477.00	424.00	371.00
22690	5226-I	10	100	532.50	479.25	426.00	372.75
78-4146-22692	5226-ISP	10	100	532.50	479.25	426.00	372.75
22682	5226SP	10	100	530.00	477.00	424.00	371.00
22702	5226WSP	10	100	534.50	481.05	427.60	374.15
22710	5227	10	100	439.00	395.10	351.20	307.30
22720	5227-I	10	100	443.00	398.70	354.40	310.10
78-4146-22750	5230	10	50	302.50	272.25	242.00	211.75
22760	5230-I	10	50	337.00	303.30	269.60	235.90
22770	5231	10	50	420.50	378.45	336.40	294.35
22780	5231-I	10	50	455.50	409.95	364.40	318.85
22790	5233	10	50	477.00	429.30	381.60	333.90
78-4146-22800	5233-I	10	50	513.00	461.70	410.40	359.10
22810	5235	10	50	508.00	457.20	406.40	355.60
22820	5235-I	10	50	543.00	488.70	434.40	380.10
22830	5236	10	50	439.00	395.10	351.20	307.30
22840	5236-I	10	50	450.00	405.00	360.00	315.00
78-4146-22850	5237	10	50	335.00	301.50	268.00	234.50
22860	5237-I	10	50	370.00	333.00	296.00	259.00
22870	5238	10	50	629.00	566.10	503.20	440.30
22880	5238-I	10	50	668.00	601.20	534.40	467.60
22890	5239	1	20	325.50	-	260.40	227.85
78-4146-22900	5240	1	20	325.50	-	260.40	227.85
22910	5241	10	50	785.50	706.95	628.40	549.85
22920	5241-I	10	50	791.00	711.90	632.80	553.70
22930	5241W	10	50	794.50	715.05	635.60	556.15
49500	5242	10	100	246.00	221.40	196.80	172.20
78-4146-49508	5242BLK	10	100	268.00	241.20	214.40	187.60
49510	5242GY	10	100	268.00	241.20	214.40	187.60
49502	5242-I	10	100	268.00	241.20	214.40	187.60
49512	5242R	10	100	268.00	241.20	214.40	187.60
49514	5242W	10	100	268.00	241.20	214.40	187.60
78-4146-22940	5243	10	50	981.50	883.35	785.20	687.05
22950	5243-I	10	50	986.00	887.40	788.80	690.20
22960	5243W	10	50	988.00	889.20	790.40	691.60
22970	5245	10	50	848.00	763.20	678.40	593.60
22980	5245-I	10	50	850.50	765.45	680.40	595.35
78-4146-22990	5245W	10	50	852.00	766.80	681.60	596.40
23042	5248-CPX	10	200	156.00	140.40	124.80	109.20
23051	5248-I-CPX	10	200	161.00	144.90	128.80	112.70
23052	5248-IX	10	200	161.50	145.35	129.20	113.05
23062	5248X	10	200	156.50	140.85	125.20	109.55
78-4146-23102	5248-4-IX	Bulk	200	161.00	-	128.80	112.70
23106	5248-4X	Bulk	200	156.00	-	124.80	109.20
49590	5251	10	50	334.00	300.60	267.20	233.80
49594	5251BLK	10	50	350.50	315.45	280.40	245.35
49595	5251GY	10	50	350.50	315.45	280.40	245.35
49592	5251-I	10	50	350.50	315.45	280.40	245.35

NOTE: These are samples only. Current prices are available from Trade Service Publications, Inc.
Courtesy: *Trade Service Publications, Inc.*

Elec. Page 168-29 B
Dec 23 1981
Cancels Nov 25 1981

LEVITON WIRING DEVICES - PER 100

DCI EDP No.	Cat. No.	Carton	Std. Pkg.	List	Bkn. Ctn.	Ctn. + Over	S. P. + Over	
78-4146-27380	8871	10	100	$ 158.50	$142.65	$ 126.80	$ 110.95	
27390	8871-3	10	100	145.50	130.95	116.40	101.85	D
27450	8875	Bulk	200	135.00	-	108.00	94.50	
27480	8880	Bulk	200	164.50	-	131.60	115.15	
27610	8902	1	10	1049.50	-	839.60	734.65	D
78-4146-27620	8903	1	50	386.50	-	309.20	270.55	D
27630	8904	1	20	773.50	-	618.80	541.45	D
27650	8906	1	25	553.50	-	442.80	387.45	D
27670	8908	1	5	2763.00	-	2210.40	1934.10	D
27680	8909	1	10	1455.50	-	1164.40	1018.85	D
78-4146-27750	8916	1	10	1640.00	-	1312.00	1148.00	D
27800	9063	10	200	151.50	136.35	121.20	106.05	
27900	9148	1	25	845.00	-	676.00	591.50	D
28052	9217	25	200	104.00	93.60	83.20	72.80	
28054	9218	25	200	214.50	193.05	171.60	150.15	
78-4146-28056	9219	1	10	545.00	-	436.00	381.50	D
28058	9220	1	10	554.50	-	443.60	388.15	D
28340	9346BG	10	250	160.50	144.45	128.40	112.35	x
28350	9346BR	10	250	213.50	192.15	170.80	149.45	x
28370	9346N1	10	250	159.00	143.10	127.20	111.30	x
78-4146-28360	9346M	10	100	97.50	87.75	78.00	68.25	
28380	9346PG	10	100	162.00	145.80	129.60	113.40	x
28390	9347BG	10	250	155.50	139.95	124.40	108.85	x
28400	9347BR	10	250	209.00	188.10	167.20	146.30	x
28420	9347N1	10	250	153.50	138.15	122.80	107.45	x
78-4146-28410	9347M	25	100	71.00	63.90	56.80	49.70	D
28430	9347PG	10	100	157.50	141.75	126.00	110.25	x
28520	9350	10	100	137.50	123.75	110.00	96.25	
28580	9700	10	250	137.00	123.30	109.60	95.90	
23590	9700-I	10	250	141.50	127.35	113.20	99.05	
78-4146-28600	9701	10	100	137.00	123.30	109.60	95.90	D
23610	9701-I	10	100	141.50	127.35	113.20	99.05	D
28613	9701K	10	100	137.00	123.30	109.60	95.90	
28615	9701-IK	10	100	141.50	127.35	113.20	99.05	
28633	9702K	10	100	131.00	117.90	104.80	91.70	
78-4146-28635	9702-IK	10	100	135.50	121.95	108.40	94.85	
28670	9708	1	100	131.00	-	104.80	91.70	
28680	9708B	10	100	131.00	117.90	104.80	91.70	
28690	9708-I	10	100	131.00	117.90	104.80	91.70	
28730	9709	10	100	321.50	289.35	257.20	225.05	
78-4146-28740	9709B	10	100	321.50	289.35	257.20	225.05	D
28750	9709-I	10	100	321.50	289.35	257.20	225.05	
28790	9715A	1	50	336.00	-	268.80	235.20	
28800	9715B	1	50	357.50	-	286.00	250.25	
28810	9715C	1	50	342.00	-	273.60	239.40	
78-4146-28830	9716A	1	50	342.00	-	273.60	239.40	
28840	9716B	1	50	363.50	-	290.80	254.45	
28850	9716C	1	50	349.00	-	279.20	244.30	

D - To Be Discontinued When Present Stock Is Exhausted.
x - Price Correction

NOTE: These are samples only. Current prices are available from Trade Service Publications, Inc.
Courtesy: *Trade Service Publications, Inc.*

Estimating Electrical Construction

Elec. Page 168-49 A Oct 19 1981 Cancels Mar 2 1981	LEVITON WIRING DEVICES - PER 100						
DCI EDP No.	Cat. No.	Car- ton	Std. Pkg.	List	Bkn. Ctn.	Ctn. + Over	S. P. + Over
78-4146-43838	86000	1	10	$ 713.00*	$ -	$ 570.40*	$ 499.10* †
43840	86001	25	250	46.00	41.40	36.80	32.20 †
43841	86001AC	25	250	220.50	198.45	176.40	154.35
43842	86001HG	10	50	252.50	227.25	202.00	176.75
43850	86003	25	250	46.00	41.40	36.80	32.20 †
78-4146-43855	86003HG	10	50	252.50	227.25	202.00	176.75
43860	86004	10	100	52.00	46.80	41.60	36.40 †
43865	86004HG	10	50	252.50	227.25	202.00	176.75
43870	86005	25	250	93.00	83.70	74.40	65.10 †
43890	86007	10	100	116.50	104.85	93.20	81.55 †
78-4146-43910	86009	25	250	93.00	83.70	74.40	65.10 †
43912	86009HG	10	20	506.50	455.85	405.20	354.55
43920	86011	10	100	138.00	124.20	110.40	96.60 †
43922	86011HG	10	20	758.50	682.65	606.80	530.95
43930	86012	10	50	200.50	180.45	160.40	140.35 †
78-4146-43932	86012HG	10	20	1011.50	910.35	809.20	708.05
43940	86013	10	100	52.00	46.80	41.60	36.40 †
43950	86014	10	100	52.00	46.80	41.60	36.40 †
43952	86014HG	10	20	282.00	253.80	225.60	197.40
43960	86016	10	100	105.50	94.95	84.40	73.85 †
78-4146-43962	86016HG	10	20	506.50	455.85	405.20	354.55
43970	86017	10	100	56.50	50.85	45.20	39.55 †
43975	86018	10	100	79.00	71.10	63.20	55.30 †
43977	86018HG	10	20	439.00	395.10	351.20	307.30
43980	86021	10	100	160.00	144.00	128.00	112.00 †
78-4146-43990	86023	10	20	397.50	357.75	318.00	278.25 †
44000	86025	10	100	167.50	150.75	134.00	117.25 †
44010	86036	10	20	461.00	414.90	368.80	322.70 †
44020	86101	25	100	96.00	86.40	76.80	67.20 †
44030	86103	25	100	96.00	86.40	76.80	67.20 †
78-4146-44040	86105	25	50	192.00	172.80	153.60	134.40 †
44044	86106	25	50	198.50	178.65	158.80	138.95 †
44046	86108	25	50	198.50	178.65	158.80	138.95 †
44050	86109	25	50	192.00	172.80	153.60	134.40 †
44060	86116	25	50	198.50	178.65	158.80	138.95 †
78-4146-44070	86301	25	100	46.00	41.40	36.80	32.20 †
44080	86303	25	100	46.00	41.40	36.80	32.20 †
44090	86305	25	100	93.00	83.70	74.40	65.10 †
44093	86306	10	100	103.00	92.70	82.40	72.10 †
44096	86308	10	100	103.00	92.70	82.40	72.10 †
78-4146-44100	86309	25	100	93.00	83.70	74.40	65.10 †
44110	86316	25	100	93.00	83.70	74.40	65.10 †
44150	87001	25	250	46.00	41.40	36.80	32.20 †
44152	87001HG	10	50	252.50	227.25	202.00	176.75
44160	87003	25	250	46.00	41.40	36.80	32.20 †
78-4146-44165	87003HG	10	50	252.50	227.25	202.00	176.75
44170	87004	10	100	54.00	48.60	43.20	37.80 †
44172	87004HG	10	50	252.50	227.25	202.00	176.75
44180	87005	25	250	95.00	85.50	76.00	66.50 †
44190	87007	10	100	93.00	83.70	74.40	65.10 †
78-4146-44200	87009	25	250	95.00	85.50	76.00	66.50 †
44202	87009HG	10	20	506.50	455.85	405.20	354.55
44210	87011	10	100	135.00	121.50	108.00	94.50 †
44212	87011HG	10	20	758.50	682.65	606.80	530.95
44220	87012	10	50	205.50	184.95	164.40	143.85 †
78-4146-44222	87012HG	10	20	1011.50	910.35	809.20	708.05
44230	87013	10	100	54.50	49.05	43.60	38.15 †
44240	87014	10	100	54.00	48.60	43.20	37.80 †
44242	87014HG	10	20	282.00	253.80	225.60	197.40
44250	87016	10	100	105.50	94.95	84.40	73.85 †
78-4146-44252	87016HG	10	20	506.50	455.85	405.20	354.55
44260	87017	10	100	54.00	48.60	43.20	37.80 †
44276	87018HG	10	20	439.00	395.10	351.20	307.30
44270	87021	10	100	160.00	144.00	128.00	112.00 †
44280	87023	10	20	398.00	358.20	318.40	278.60 †

* - Per 100 Boxes

NOTE: These are samples only. Current prices are available from Trade Service Publications, Inc.
Courtesy: *Trade Service Publications, Inc.*

SLATER WIRING DEVICES - PER 100

DCI EDP No.	Cat. No.	Car-ton	Std. Pkg.	List	Bkn. Ctn.	Ctn. + Over	S. P. + Over	
78-5804-41858	S94071	25	250	$ 46.00	$ 41.40	$ 36.80	$ 32.20	†
41808	S94072	10	100	93.00	83.70	74.40	65.10	†
41805	S94073	10	100	138.00	124.20	110.40	96.60	†
41862	S94074	10	100	200.50	180.45	160.40	140.35	†
41863	S94075	10	100	397.50	357.75	318.00	278.25	†
78-5804-41797	S94076	10	100	461.00	414.90	368.80	322.70	†
41865	S94091	25	250	52.00	46.80	41.60	36.40	†
41866	S94101	25	250	46.00	41.40	36.80	32.20	†
41820	S94102	10	100	105.50	94.95	84.40	73.85	†
41825	S94121	25	250	52.00	46.80	41.60	36.40	†
78-5804-41856	S94122	10	100	167.50	150.75	134.00	117.25	†
41867	S94181	25	250	56.50	50.85	45.20	39.55	†
41845	S94532	10	100	93.00	83.70	74.40	65.10	†
41855	S94543	10	100	160.00	144.00	128.00	112.00	†
41853	S94572	10	100	115.50	103.95	92.40	80.85	†
78-5804-41920	SH91071	10	50	-	214.20	188.90	166.90	
41980	SH91072	10	100	-	416.60	367.20	324.50	
41921	SH91091	10	50	-	214.20	188.90	166.90	
41922	SH91101	10	50	-	214.20	188.90	116.90	
42000	SH91102	10	100	-	416.60	367.20	324.50	
78-5804-42050	SH91532	10	100	-	416.60	367.20	324.50	
41924	SH92071	10	50	-	240.40	211.90	187.20	
41982	SH92072	10	100	-	482.10	425.00	375.55	
41925	SH92091	10	50	-	240.40	211.90	187.20	
41926	SH92101	10	50	-	240.40	211.90	187.20	
78-5804-42002	SH92102	10	100	-	482.10	425.00	375.55	
42052	SH92532	10	100	-	482.10	425.00	375.55	
41928	SH93071	10	50	-	253.00	221.40	197.30	
41984	SH93072	10	100	-	507.50	444.10	395.85	
41929	SH93091	10	50	-	253.00	221.40	197.30	
78-5804-41930	SH93101	10	50	-	253.00	221.40	197.30	
42004	SH93102	10	100	-	507.50	444.10	395.85	
42054	SH93532	10	100	-	507.50	444.10	395.85	
41932	SH94071	10	50	-	253.00	221.40	197.30	
41986	SH94072	10	100	-	507.50	444.10	196.85	
78-5804-41933	SH94091	10	50	-	253.00	221.40	197.30	
41934	SH94101	10	50	-	253.00	221.40	197.30	
42006	SH94102	10	100	-	507.50	444.10	395.85	
42056	SH94532	10	100	-	507.50	444.10	395.85	
41936	SH95071	10	50	-	253.00	221.40	197.30	
78-5804-41988	SH95072	10	100	-	507.50	444.10	395.85	
41937	SH95091	10	50	-	253.00	221.40	197.30	
41938	SH95101	10	50	-	253.00	221.40	197.30	
42008	SH95102	10	100	-	507.50	444.10	395.85	
42058	SH95532	10	100	-	507.50	444.10	395.85	
78-5804-13130	SIR-DT	-	1	9200.00	--	-	6440.00	
-	SIR-EX-BR	10	50	198.50	178.65	158.80	138.95	
-	SIR-EX-IV	10	50	198.50	178.65	158.80	138.95	
42736	SIR-F-SPG	1	10	4475.00	-	3580.00	3132.50	
42723	SIR-15F-BR	1	10	4475.00	-	3580.00	3132.50	
78-5804-42724	SIR-15F-BR-HG	1	10	5925.00	-	4740.00	4147.50	
42725	SIR-15F-IV	1	10	4475.00	-	3580.00	3132.50	
42726	SIR-15F-IV-HG	1	10	5925.00	-	4740.00	4147.50	
42729	SIR-15F-WH	1	10	4475.00	-	3580.00	3132.50	
42730	SIR-15F-WH-HG	1	10	5925.00	-	4740.00	4147.50	
78-5804- -	SIR-15-HH	1	5	9000.00	-	7200.00	6300.00	x
42737	SIR-20F-BR	1	10	4675.00	-	3740.00	3272.50	
42738	SIR-20F-BR-HG	1	10	6125.00	-	4900.00	4287.50	
42739	SIR-20F-IV	1	10	4675.00	-	3740.00	3272.50	
42740	SIR-20F-IV-HG	1	10	6125.00	-	4900.00	4287.50	
78-5804-42743	SIR-20F-WH	1	10	4675.00	-	3740.00	3272.50	
42744	SIR-20F-WH-HG	1	10	6125.00	-	4900.00	4287.50	†
41875	SJ91071	10	100	61.00	54.90	48.80	42.70	†
41876	SJ91072	10	100	122.00	109.80	97.60	85.40	†
41901	SJ91073	10	100	233.00	209.70	186.40	163.10	†
78-5804-41902	SJ91091	10	100	68.50	61.65	54.80	47.95	†
41877	SJ91101	10	100	61.00	54.90	48.80	42.70	†
41903	SJ91121	10	100	68.50	61.65	54.80	47.95	†

x - Package Change

NOTE: These are samples only. Current prices are available from Trade Service Publications, Inc.
Courtesy: *Trade Service Publications, Inc.*

Elec. Page 175-17 A
Jan 13 1982
Cancels Sep 28 1981

NUTONE RESIDENTIAL SECURITY SYSTEMS - EACH

DCI EDP No.	Cat. No.	Description	Std Pkg.	Retail	Dealer
		WIRELESS INTRUDER DETECTION/ALARM SYSTEM			⊕
		Includes Master Control, 2-S2291, 2-S2279			
		1 Power Plug + Mounting Hardware			
78-4891-10523	S2255F1	Frequency 1	1	$379.60	$253.07
10524	S2255F2	" 2	1	379.60	253.07
10525	S2255F3	" 3	1	379.60	253.07
10526	S2255F4	" 4	1	379.60	253.07
		MASTER CONTROL UNITS ONLY			
		Intruder/Fire Alarm System			
78-4891-10763	SA2300	Finish Kit	1	343.20	228.80
10490	SA22/23R	Recessed Housing	1	68.90	45.93
		Intruder Alarm System			
78-4891-10521	SA2250	Finish Kit	1	293.80	195.87
10490	SA22/23R	Recessed Housing	1	68.90	45.93
10516	S22/23SF	Surface Housing	1	48.50	32.33
		ALARMS ONLY			
78-4891-10435	S205	Automatic Alarm Shutdown	1	59.90	39.93 ⊕
10780	S2330	Ditto for S2250 and S2300	1	36.80	24.53
10786	S2332	Outside Electronic Siren	1	96.90	64.60
10795	SA2335	Inside Alarm Horn	3	33.60	22.40
10815	S2340	Inside/Outside Alarm Bell	1	55.30	36.87
78-4891-10830	S2345	Tamper Proof Outside Alarm Bell	1	110.80	73.87
10840	S2349	W.P. Wiring Box for S2340	1	18.70	12.47
10905	S2379	Auxiliary Relay	1	41.70	27.80
		HEAT AND SMOKE DETECTORS			
78-4891-10140	S120	Fire-Heat Detector - 135°	12	7.50	5.00 ⊕
10145	S121	Fire-Heat Detector - 200°	12	7.50	5.00 ⊕
10147	S122	Fire-Heat Detector - 135°	6	14.80	9.87
10149	S123	Fire-Heat Detector - 200°	6	14.80	9.87
10152	SA124	Heat Detector - 135°	12	6.40	4.27
78-4891-10154	SA125	Heat Detector - 200°	12	6.40	4.27
10248	S240	Supervisory Relay	1	21.20	14.13
10256	S246	Smoke Detector	1	114.00	76.00
10261	S246H	Heat/Smoke Detector	1	118.00	78.67
		ACCESSORIES			
78-4891-10900	S2378	Auxilary Power Supply	1	48.80	32.53
		INTRUDER DETECTORS			
78-4891-10511	S2244	Audio - Perimeter or Interior	1	110.10	73.40
-	S2246	Ultrasonic Motion	1	216.00	144.00 N
10514	S2248	Break Glass Perimeter or Interior	1	43.40	28.93
10531	SA2260	Recessed Plunger	24	3.50	2.33 D
10532	SB2260	Roller/Plunger Detector	25	8.00	5.33 D
10535	S2261	Surface Plunger Detector	24	5.30	3.53 D
78-4891-10541	SA2262	Foil Contact Switch	24	6.10	4.07
10545	S2263	Window Foil Tape - 320' Roll	3	11.80	7.87
10551	S2264	Foil Connector Blocks	24	.70	.47
10555	S2265	Foil Sealer - 4 oz. Container	4	5.60	3.73
10560	S2266	Recessed Magnetic Detector	24	5.30	3.53
78-4891-10565	S2267	Wide Gap Surface Detector	6	37.60	25.07
10570	S2268	Recessed Magnetic Detector	24	4.50	3.00
10573	S2268A	Steel Door Adapter for S2268	6	2.60	1.73
10575	S2268B	Adaptor	6	2.60	1.73
10580	S2269	Surface Magnetic Detector	24	6.00	4.00
78-4891-10581	S2269D	Terminal Cover	24	.70	.47
10600	S2271	Floor Mat Detector	1	36.90	24.60
10605	S2271-25	Floor Mat Detector 25' Roll	1	100.10	66.73
10621	S2273	Recessed Plunger Detector	24	5.00	3.33 D
10622	SA2273	Recessed Plunger Detector	25	8.00	5.33 N
10624	S2274	Tamper Switch	3	4.90	3.27 A

N - New Item
⊕ - Item No Longer Listed in Mfr.'s Price Sheet

NOTE: These are samples only. Current prices are available from Trade Service Publications, Inc.
Courtesy: *Trade Service Publications, Inc.*

Sample Price List
Figure 9-2

Elec Page 280-25
Mar 31 1982
Cancels Apr 16 1980

Mfr's. Zone Western

COPPER COATED PRODUCTS - PER 100

Size	Std. Pkg.	Bkn. Pkg.	S. P. to 99
GROUND RODS			↓
3/8" x 5'	10	$ 348.00	$ 292.00
3/8 x 6	10	418.00	351.00
3/8 x 8	10	556.00	468.00
1/2 x 5	5	524.00	440.00
1/2 x 6	5	618.00	520.00
1/2 x 8	5	804.00	675.00
1/2 x 10	5	990.00	832.00
5/8 x 5	5	626.00	526.00
5/8 x 6	5	740.00	622.00
5/8 x 8	5	970.00	815.00
5/8 x 9	5	1080.00	908.00
5/8 x 10	5	1200.00	1007.00
5/8 x 12	5	1500.00	1260.00
5/8 x 15	5	1855.00	1558.00
3/4 x 8	3	1531.00	1286.00
3/4 x 10	3	1891.00	1590.00
3/4 x 12	3	2385.00	2003.00
3/4 x 15	3	2988.00	2510.00
1 x 10	2	4050.00	3402.00
1 x 20	1	-	6946.00
GROUND RODS WITH 18" COPPER PIGTAILS			↓
3/8 x 5'	10	480.00	403.00
3/8 x 6	10	558.00	468.00
1/2 x 5	5	888.00	746.00
SECTIONAL GROUND RODS			↓
1/2" x 10'	5	1065.00	895.00
5/8 x 8	5	998.00	839.00
5/8 x 10	5	1234.00	1036.00
3/4 x 10	3	1948.00	1636.00
1 x 10	2	4173.00	3505.00

Size	Std. Pkg.	1 to 9	10 + Over
COUPLINGS SECTIONAL GROUND RODS			↑
1/2"	25	$325.00	$260.00
5/8	25	424.50	339.60
3/4	25	929.80	503.00
1	25	2391.60	1913.00
STUD BOLTS FOR SECTIONAL GROUND RODS			↑
1/2"	10	223.50	178.80
5/8	10	225.60	180.50
3/4	10	235.60	188.50
1	10	383.50	306.80

GROUND ROD CLAMPS

	Bkn. Pkg.	S. P. + Over
With Safety Screw or Hex Head Bolt		↑
1/2" 100	$225.20	$160.90
5/8 100	265.00	189.30
3/4 100	321.80	229.90
1 100	514.60	367.60

NO CASH DISCOUNT

NOTE: These are samples only. Current prices are available from Trade Service Publications, Inc.
Courtesy: *Trade Service Publications, Inc.*

Sample Price List (continued)
Figure 9-2

Chapter 10

Labor Units

There is no such thing as the perfect labor unit. All must be tailored to fit job conditions and installation procedures.

The past productivity of your own crews on similar jobs is the best guide to the labor that will be needed on your next job. Labor units that have been tested and refined on many jobs will be the most reliable figures you can find. Still, you always have to adjust labor units to fit the task at hand.

Labor units vary from contractor to contractor. That's because the needs and methods of contractors vary. Such things as size of operations, equipment and tooling, purchasing habits, job experience, and field supervision affect labor productivity.

Labor units are expressed in manhours. A set of labor tables for electrical work is included at the end of this chapter. Use these figures to supplement your own data until you develop figures you know work better.

Acquiring and Evaluating Labor Units

It would be almost impossible to list labor units for every material and equipment item. Those listed in this chapter make a good benchmark from which to begin. Each time a labor unit is applied to the estimate, you have to ad-

just the benchmark to fit job conditions.

Start a log to record your own labor units. Divide the log into subsections so that material or equipment categories can be easily found and identified. Each time you have to develop a new installation time, list that new unit in the log for future reference. Make another entry when the job is complete and you know what the actual labor time was. Review and update your labor unit estimates occasionally so they become more accurate.

Don't be afraid to ask others for information about certain field installations. No estimator has all the answers. Stay current on changes taking place in the industry. Many of these changes can help you become more competitive and do a better job. New material will need to be evaluated; improved material will need re-evaluation. New and better tools might reduce the installation time at the job site. Better electrical equipment could lead to installation savings, also.

Equipment Assembly

Be aware of excessive equipment assembling at the job site. Some manufacturers try to save money by shipping unassembled equipment. This is often the case with lighting fixtures, switchboards, panelboards and motor control units.

Most labor units don't allow for the extra time it takes to assemble equipment that normally arrives completely assembled. And usually the extra work isn't noticed until after the job has been bid. Most experienced estimators have learned the hard way which manufacturers are likely to cut corners.

Here are some items that may arrive assembled or unassembled:

Fixtures
- Socket assemblies
- Protective devices
- Closure and end plates
- Internal wiring
- Ballasts
- Diffuser components

Switchboards
- Circuit breakers
- Fuses
- Bus connections
- Nameplates
- Other cabinet components
- Grounding parts
- Control, metering & protective devices

Panelboards
- Circuit breakers
- Nameplates
- Cover hardware
- Bus hardware
- Grounding components
- Other panelboard components

Adverse Job Site Conditions

Conditions on some jobs complicate installation procedures. You should be very sensitive to both obvious and less obvious situations that may require extra labor. The following are typical examples of adverse job site conditions.

Weather: Cold weather and snow removal reduce efficiency. During the winter months your crews are going to take more time on almost every type job, especially when the building offers little protection from the elements.

Extremely hot weather can also affect production. Sometimes split shifts are necessary—starting early, taking a break during the hottest hours, and returning later to complete the day.

Rain affects outside work, especially underground construction. Not only will more time be needed for each task, but additional work such as pumping, mucking and cleaning may be necessary.

Wind also can cause problems on outside work. Tower and other elevated work becomes especially difficult.

On-site congestion: Jobs located in downtown areas usually have limited work space. That hampers unloading of materials and equipment. This wastes manhours that should be used on the installation.

Jobs in remote locations require additional time to move tools, materials and equipment to where they're needed. Too much handling can cause damage and loss.

Parking problems or lack of parking space reduces productive time. And lack of material storage space can lead to material shortages. This means additional trips to the suppliers and wasted manhours.

Construction schedule: A construction schedule that calls for completion of one phase before starting on the next phase can result in additional manhours. Such schedules usually divide a project into smaller jobs, requiring additional material deliveries, additional layout of the work and additional coordination with the other trades.

The job may require the complete removal of material, equipment and tools between the end of one phase and the start of the next. And the down time between the scheduled phases may be so long that the workers and supervisors will be changed. The result is additional manhours.

Job location: Travel distance is an important consideration. Jobs outside your normal working area may create problems finding qualified help. Chances are the good help in the area is already on someone else's payroll. That's why labor productivity may be lower if you can't use your regular crews.

Material deliveries may become a problem, especially when small amounts of will-call items are needed. It's almost impossible to purchase the exact amount of material that will be required to complete the installation. And deliveries take longer and can't be made as often. When material supplies become a problem, production suffers.

Remote job sites reduce management access. In general, the level and quality of production correspond with management participation in overall field supervision. Regular management review of the work in progress helps control costs. If management can't check the job regularly, production will be lower.

Overtime: Overtime jobs are generally less productive. Worker efficiency decreases as extra hours are worked. Working six 8-hour days has little effect on worker performance. But working more than eight hours a day lowers production efficiency. Extra hours result in fatigue, boredom, and less production.

Sometimes it's better to work overtime to complete a job that requires a lot of set-up time. This is a management decision, not the estimator's. If you know that overtime is necessary, include it in your labor estimate.

Minor change orders: These disrupt the normal progress of the job. Either the planned layout must be changed or the installed work must be altered. With a little experience and good job records, you can identify projects that may require minor changes.

Changes may create material shortages on the job if material purchased for the original work is used on the change. Some material may even become surplus because of the change. And job progress may be slowed by new material requirements. The new item may not be readily available, and it may take time to locate and deliver.

Most change orders require pricing and approval before the change is made. This, too, can add to production problems.

Many workers don't like to redo their work. Frequent changes to work already in place can affect morale and, in turn, affect the whole job.

Scaffold work: Working on scaffolding restricts production. Getting to the work area is more difficult. The higher the scaffold, the more time needed for installation.

Usually additional help is needed at ground level to assist the workers on the scaffold. Materials must be prepared and hoisted to the working level.

The ground around rolling scaffolds must be clear so that the scaffold can be moved. Workers on the ground can clear the area of small amounts of rubble, but large deposits of material and equipment can cause problems. Sometimes scaffolds must be partly dismantled for movement from one area to another or around beams, girders, ducting, draft curtains, piping or other obstructions.

Exposed systems: Exposed systems must be plumb and level. Conduits must be offset at each point of termination and around obstructions. Additional conduit bodies or junction boxes may be required to work around obstructions. Concentric bending of offsets and elbows in multiple runs of conduit may also be necessary. And additional conduit and wire is usually needed in exposed wiring systems.

Special hanging assemblies may be required to provide support for groups of conduit. Special suspension brackets may be necessary to install panelboards and distribution equipment.You may even need special suspension for lighting fixtures to maintain a specified mounting height.

Job design: Inadequate plans and specifications can create big problems, especially at the job site. Poorly designed construction documents make your job and the installation very difficult.

Plans and specifications that are unclear or have conflicting details do not produce a good job. When confusion is built in to the design of the job, the actual construction becomes extremely difficult.

Reducing plan size to save paper also creates problems. Reduced plans are usually hard to read. The problem is compounded if the installation must be made under poor conditions. Some designers provide clear details to ensure that the installation is done properly.

Too many vague details can clutter up a set of plans. It may be to your advantage to work only on plans drawn with clear and helpful details.

Labor supply: Be aware of the labor available to complete the job. When there's a shortage of skilled labor, the manhours in the estimate must be adjusted accordingly.

If your company has a heavy labor attrition rate, factor the total labor or increase your labor units.

Unusual requirements: Study unusual job requirements to determine their effect on the estimated labor. Specialists may be needed for difficult or unusual installations. Many companies hire others for special services such as high voltage cable splicing, testing or calibrating. And it's common to call in communications technicians to install sound and video systems, fire alarm systems, burglar alarm systems, and other special or unusual setups.

Some companies subcontract with specialists

for the installation of street lighting or parking area lighting. Usually it's hard to beat the price of a specialist.

Addenda: In most cases, addenda are issued to correct mistakes made by the design team. A well-designed job seldom requires an addendum.

An addendum can affect the original job plan even if that plan is not changed. Most architects don't provide a corrected set of construction plans and specifications when the contract is let. They depend on the contractor to correct the plans and specs with notations referring to the addendum. When architects issue a series of long addenda, the problem is even more serious.

You'll often bid a job after seeing only the plans and specs that apply to you. Addenda issued to the general contractor may not be passed on to all the subs. Many times the general doesn't foresee how the change will affect other work and doesn't realize certain subs should be notified.

As construction progresses, some subcontractors may install work that should have been changed by an addendum. This can affect others when the work is corrected.

Restrictive job site conditions: Some sites are so confined that all trades have difficulty installing their work. Poor coordination by the general contractor will aggravate this problem. Too many trades working in a tight space reduces efficiency.

Restrictive conditions can reduce on-site storage space. This leads to material shortages and delivery problems.

Job site security regulations can reduce productivity. Workers subject to security checks may be delayed in their work. Security clearances that have to be renewed frequently increase job overhead.

Hazardous conditions require extra safety measures. These can be restrictive and costly.

Some plants and factories can't be shut down for alterations or additions. This may require off-hour work which can cause fatigue and lead to poor job performance.

Public safety: Every electrical contractor is responsible for public safety. Your work has to comply with safety codes and regulations. These are the minimum standards. An error in design is no excuse for failure to comply with the code.

Public safety may mean any or all of the following: warning signs, barricades, fencing, isolation vaults, isolating underground systems, metal clad equipment, grounding systems, heavily insulated materials, ground fault interrupters, locking systems and alarm systems.

Lack of utilities: Water and electrical service may not be available on the job site until late in the job. Portable generators can take up some of the slack. But they don't let you test electrical systems as they are installed. And additional labor must be spent for trouble shooting.

Beneficial occupancy: The contract may allow for "beneficial occupancy" by the owner. That means the owner or his tenant move in before the work is finished. This can create problems. It's hard to work around the intended occupants of a building. They probably want certain areas completed first, even though this conflicts with your construction schedule.

An occupied job reduces storage space. Materials are either dumped into lockup boxes or piled up in the open areas. Material damage and loss can result.

Inspections: Most jobs require inspections. These are usually done by the local building official or by the customer's inspector. Sometimes the architect is responsible for inspections to ensure compliance with the contract. Sometimes two or more inspectors with unrelated authority will be inspecting the project. Problems multiply on overly inspected jobs.

Lengthy jobs: Some jobs seem to drag on forever. Each contractor has an idea of how long a job should take. Most contractors don't want to be involved on a project for more than a year. Others take on jobs that run up to five years.

Long range jobs are not necessarily big jobs. They can be moderate in size and be spread out over a long, segmented schedule.

Non-productive Burdens

Your labor estimate always includes non-productive labor time. This is in addition to the actual time it takes to perform a specific task. For example, a time study might show that an average of 0.25 hours is required to complete a task. But few workers spend all of their work time actually doing productive work. There will always be interruptions. These must be added to the actual productive labor time.

Non-productive labor can add 20% or more

to actual job time. A job that takes 0.25 hours multiplied by 1.20 (actual job time plus 20%) becomes 0.30 hours. On most jobs and for most electricians, you can figure about 50 minutes of each hour will be productive. That's a good standard.

The following are non-production burdens common to most jobs. Undoubtedly there are others. Try to spot unusual non-productive labor situations that increase manhour requirements.

After looking over the list that follows you'll begin to appreciate the need for applying the right amount of labor hours to each individual work item. You'll also see why material can be estimated more accurately than labor. So many variables affect every labor estimate. Don't underestimate labor units. Electrical contracting involves much more than just installing the material.

Material, tool and equipment receiving and storage: Labor is needed to unload, count, check and record shipments at the job site. Packing slips and freight bills must also be checked.

The shipment must be moved from the point of delivery to either the installation point or a storage area. Material may have to be moved from one storage area to another. And it must be moved to the work area and unpacked, assembled and installed.

Material ordering: Few jobs go from start to finish without needing some materials that were not anticipated. Ordering this material takes time. Hundreds of minor items are needed for the many installations. For example, screws, lugs, tape, beam clamps, metal channel and fittings, hanger rods, reducing washers and bushings, pulling compounds, cutting oil, rags and conduit seals don't show up on the take-off pricing sheets. But every job you have will need some of these.

Many contractors don't ship all of the material to the job at one time. Instead, partial shipments are sent to start the job. The rest of the material must be requested as needed.

Tools and equipment for installing the work must also be ordered as needed. Often parts are needed to keep equipment operating. Such items as hacksaw blades, drill bits, hole saws, chalk line, crayons, pencils, paper, forms, benders, threading equipment, drill motors, ladders, extension cords, and cord wyes all must be ordered.

Storage sheds are usually sent on schedule, but lockup boxes, lockup cabinets and storage shelving are usually ordered as needed.

First aid supplies and safety equipment must be ordered.

Plan and specification review: Those installing the electrical work must have time to study the plans, specifications and structural design. Embedded items must be located in the structure, all conduit routing must be planned, and hookup to the equipment must be checked and coordinated with the other trades that need hookups. Such items as conduit, wire size, equipment requirements and space for lighting-distribution panels and cabinets must be checked and verified before the work can begin.

Conduit routing is most important at the start of installations. Many times there will be obstacles that the design team missed.

Someone has to locate and size chases and sleeves to be sure there is good access for future installation. Chases may be simple blockouts for conduit stubs or they may be detailed blockouts for lighting fixtures, panelboards, busways and wireways.

Your crew leader will review the plans and installation requirements and then decide on the makeup and size of his crew.

Material schedules and tools must be verified. Tools must be on hand to install the material.

A poorly planned job usually has poor labor productivity. One mistake leads to others, and labor hours are wasted.

Timekeeping: Someone must account for the time of each worker during the day. Accurate time reporting is essential for doing the weekly payroll, job costing, and tabulating the hours used for each coded task. Be sure to record a worker's classification, wage rate, deductions, fringe benefits and the accumulated hours for the reporting period.

Safety meetings: Job site safety meetings delay production. But they also reduce accidents and keep insurance rates low. All workers must attend. Usually the job foreman conducts the meeting. On large jobs the general contractor may conduct the meeting for all trades.

Breaks: Coffee and rest breaks are common throughout the construction industry. As a job proceeds, you can expect the breaks to become longer and more frequent. Sometimes they include a meal, or a trip to a coffee shop. Some jobs start the day with a coffee break, followed by another at mid-morning, then lunch, another break in mid-afternoon and an early quitting time. You're not going to stay in business long under those conditions. But you, as the estimator, have to include all non-productive time that can be expected on the job.

Inspection tours: Each time the inspector drops by, the job leader is interrupted for the duration of the visit. The more inspection agencies, the more visits and tours. The bigger the job, the longer the tours.

Many hours can be spent trying to work out a problem with the inspector. No one likes doing work over again, especially to satisfy some questionable interpretation of the code.

Coordination meetings: Many general contractors require that all subcontractors participate in weekly job site coordination meetings. The general contractor wants information from each sub on material ordering, deliveries, change orders, manpower, and task completion. The meetings are held in the hope of avoiding subcontractor delays, but they can take hours from job production.

Supervision: Your supervisor will spend many hours planning the work and trying to meet the general contractor's schedule. Supervision time is required for layout of the work, work assignments and instructions. Supervision continues until the job is finished. At the beginning of each day the supervisor must assign the work and add additional tasks as progress is made.

Payments: Most job contracts provide for progressive monthly payments to the contractor for completed work. The amount of work installed determines the payment due. On some jobs the owner's representative must approve a percentage of certain line items to qualify for a progress payment. Usually the amount of completion is established by the percentage of line items installed. The supervisor and the owner's representative make this decision. The progress billing usually is settled monthly, but can take longer to resolve.

Labor Savings

Any electrician who hangs the same fixture 50 times on one job is going to install the last fixture much faster than the first. There is a natural learning curve that applies on many types of electrical work. You will want to reduce the estimated time to allow for higher productivity on repetitive tasks. The following are some areas where labor savings are common:

Multiple conduit runs between two points: An example would be installing four 2-inch conduits parallel between two pull boxes 100 feet apart, compared to four separate runs of 2-inch conduit of 100 feet, routed in different directions. The layout time would be reduced, and the actual installation time would be similar to a production line task. Material handling would be more confined and easier to control. Installation set-up time would be much less than if it were spread over 400 feet.

Multiple wire pulling: Continuing with the previous example, the next step would be to install four sets of feeder cables of equal size in the conduits between the two pull boxes. This should take less labor than installing one set of feeders, pulled in 100-foot sections over a total of 400 feet.

There would only be one cable reel setup at one location for pulling the cable through each conduit, compared to at least two cable reel setups under the best conditions. Splicing time would also be reduced because all of the splicing could be done in just two places instead of four.

Oversized conduits: Wire pulling for oversized conduits is usually much easier. The savings, however, would be slight. All of the individual work items would still have to be performed; only the actual pulling would be easier.

Multiple conductors: Most conduit uses four conductors. But you can save labor by installing twelve conductors in conduit. The total footage of wire pulled will be three times the average pull. Savings are slight, however, because each wire must be identified, and normal color coding of the wire covers four conductors. Therefore, when there are three of each color, it may be difficult to identify the separate sets.

Economy of scale: In large residential developments, where home designs vary only slightly, labor time will usually be reduced. Handling the same work 50 times on one job almost always takes less time than doing the work once on 50 jobs. The electricians become very efficient at roughing-in and trimming each house. Payment on a piece work basis also reduces labor costs.

There are economies of scale on chain store projects, even when the work is done at widely scattered locations. The same tradesmen build several stores with similar layouts. Construction moves quickly because the workers are familiar with the design and building methods.

High-rise buildings are another example. Each floor probably has the same job requirements, materials, and construction sequence.

Similar floor plans make layout and preparation easier and much faster as the job progresses. Material control is better because types and quantities are known through experience. And the output of each worker can be monitored on each job. Workers can be assigned to the tasks they handle best. Slower workers can be replaced before too much time is lost.

Modular wiring systems for power, lighting and communications installed in high-rise buildings are also repetitive work items. After the first few floors have been completed, the installation pace should pick up.

First, the installers lay out the modular systems. They check the approved shop drawings, catalog data, and architectural room details. Then they unpack the modular systems and sort out the parts. Once this is done the installation can begin.

The manufacturer's representative may be present to assist in the initial layout and to explain the installation in detail. The first couple of floors will determine the material and labor needed for the rest of the project.

Tasks that can be easily monitored: Many installation tasks can be monitored for peak production.

• Large quantities of the same type of lighting fixtures in large office, warehouse or assembly areas where there are few obstructions. As the installation progresses, the time to install the fixture should decrease. The workers should become better organized and should reach peak production after just a few hours.

• Branch circuit wiring of typical conduit systems repeated within a large area. Workers become quicker as the job proceeds.

• Typical feeder wire installations from a common distribution location. The feeder wire reels are set on reel jacks at a common location. Because more than one feeder can be pulled from this setup, the task becomes easier.

Subsequent pulls become faster as each worker is assigned a particular pulling task. The time for reel setup is also reduced with each pull.

• Large quantities of wiring devices installed in laboratories, assembly areas, repair shops, service centers, etc. Once the first installation is completed, the worker knows how the rest should be done.

• Special jobs, such as street lighting, where the installation is reduced to work segments and each segment is repeated many times. First the layout, then the excavation, then the construction of the concrete pole bases, then the trenching and conduit installation between the concrete pole bases, then the erecting of the assembled street lighting standards, and finally the wire pulling and hookup. Each is scheduled separately and is completed before the next segment starts.

Competition among work crews: A little friendly rivalry among work crews can increase production. Workers must have the same access to tools, equipment and materials, and have similar skills. With the right supervision you can encourage the crews to outperform each other.

Soft wiring systems: Soft wiring systems are common in large offices. Checking installation and labor efficiency in large soft wiring jobs is the same as in modular wiring systems. With a little experience the crew will become very productive and efficient. Identifying the switching cables and connectors and the system's general layout are time consuming at first. But once the crew has mastered the system's layout, production time should decrease.

Soft wiring systems are very basic and don't require much training or installation skill. All that's needed is an accurate set of drawings and clear identification of all system materials.

Pre-assembled lighting fixtures: When many of the same fluorescent lighting fixtures are to be installed in large areas, they should be pre-assembled at the factory. Factory labor is, generally, much less expensive than job site labor. The more complete the lighting equipment, the less field labor will be used.

The factory can install a flexible conduit "whip" complete with connectors and wire for connection to the conduit system j-box. They can also put the lamps in the fixtures. This will save a lot of time at the site.

The length of the whip can be included in the fixture order. So can wire, lamps, and lighting control switches.

Orders for fluorescent lighting fixtures can include plug-in receptacles for soft wiring systems. When the soft wiring manufacturer has been chosen, the plug-in receptacle can be specified with the order.

Pre-cut feeder cable: You can save on labor by ordering long runs of feeder cable in cut lengths. For example: 230 feet of 2-inch conduit with 4 No. 1/0 THW, CU cable between the Main Switchboard (MSB) and the Distribution Panel (DP 1).

Additional cable will be needed for makeup in MSB and DP 1. A reasonable amount would be 10 feet per cable at each panel. The length of 1/0 THW, CU on the pricing sheet should include 1,000 feet of feeder cable.

Assuming that the field measurement confirms the feeder length of 1,000 feet, the purchase order for the cable should be as follows: 1,000 feet 1/0 THW, CU, (4 pieces at 250 feet each).

There will probably be a cutting charge of a few dollars. But that would be much less than having the workers measure and cut the cable into four 250-foot pieces.

The cable should be delivered on 4 wire reels which can be set up on dispensing jacks for pulling directly into the conduit. This reduces handling and makes cable pulling more efficient.

Example:

4 No. 1/0 THW, cu	250'/Each
4 No. 1/0 THW, cu	175'/Each
4 No. 1/0 THW, cu	175'/Each
4 No. 1/0 THW, cu	150'/Each
4 No. 1/0 THW, cu	150'/Each
4 No. 1/0 THW, cu	100'/Each

The purchase order would read:

4,000 ft. No. 1/0 THW, cu, (4 pcs. at 1,000 ft. each)

Assuming that all six runs of 1/0 feeder cable terminate at MSB, only one setup would be required at MSB. All feeders could be pulled from that setup, saving manhours.

There's another advantage to including feeder size and type in the order. When each feeder run is pulled, the cable can be cut to fit the connection at MSB and DP 1. That should reduce wasted cable. Even the best workers make mistakes when measuring cable runs for material orders. If the measurement is wrong and the cable is too short, it may have to be replaced. If the short piece of cable was included in a combined length, it might fit if the other cables are cut to fit at the point of connection.

Adjusting Labor Units

Be careful when adjusting labor units. Benchmark labor units should be based on standard contract documents, skilled workers, good supervision, working conditions, climate and site conditions, adequate materials, reasonable inspection authority, on-time deliveries, and good coordination with other trades.

Study the entire project before making any adjustments. If the installation seems to be average, but site location might be a problem for deliveries and storage, you can adjust the total labor by a few percentage points.

Adjustments may be necessary for difficult installations. Some of these difficulties are limited access, adverse climatic exposure, above average heights, exposed work, off-hour or shift work, too much overtime work, distance to the job site, congested areas, public safety, and traffic control. Sometimes you'll decide to reduce manhour requirements because conditions are closer to ideal than your benchmark labor figures assume.

Guessing how much good supervision will save can be tricky. A supervisor is usually assigned to a problem job that demands more attention. The easier job often turns out to be the real problem because it was assumed that no supervision was needed.

Adjusting the labor unit downward, if done wisely, may help get the contract and still provide enough manhours to complete the job. Favorable conditions can result in real savings. Sometimes special tools and equipment can increase installation efficiency and save manhours.

Trenching is a good example. The right trenching equipment must be used. A wheel trencher will dig a good, uniform trench that makes backfill easy to work with. The ground must be stable yet not too rocky or wet or have too much clay. When the machine digs a uniform trench, conduit can be installed more easily, encasement material can be placed more uniformly and the backfilling and compaction operation will be easier to control.

If trenching is difficult and a wheel trencher cannot be used, a backhoe may be needed. The work won't be as uniform and will require more manhours.

The tools and equipment for installing rigid conduit can vary a great deal. This is true when the specifications restrict the use of threadless couplings and connectors and require all joints to be reamed and threaded with approved tools.

Threading equipment can range from a hand die and a vise stand to an automatic threading machine. Hand threading takes longer, especially as the pipe size increases. The automatic machine usually has a cutter, reamer, adjustable dies and automatic cutting oil applicator. These machines are often used on projects requiring rigid metallic conduit.

Conduit bending machines can save many manhours. These machines can make a stub up in the conduit to the point of termination, eliminating the need for elbows, couplings and pipe nipples. Usually they are used on conduit larger than 1 inch. Ratchet mechanical benders are great on the smaller conduit. The hydraulic and mechanical benders provide uniform bending for exposed work where appearance is important.

Wire and cable pulling equipment can save time too. Spring steel fishtapes are used to pull branch circuit and feeder wire through the conduit systems. Spring leaders attached to the steel fishtape help push it through the conduit. Blower and vacuum units are available to send a pull string through the conduit. A heavier line is generally pulled along with the string, then the wire or cable is pulled in. Wire lubricants are available for reducing friction during pulling.

Manually operated winches are available for wire and cable pulling. Most winches have a flexible steel cable on a small drum that will pull several hundred pounds.

Power-driven pulling machines are needed for large, heavy pulls. A heavy rope is pulled into the conduit and attached to the cable. The rope is wound around the capstan on the power unit but slips on the capstan, thus regulating the pulling speed.

You can put a cable pulling winch on one of your utility trucks if you do much outdoor pulling. The winch should have a cable drum with a length of steel cable. For rope pulling, the winch drives a capstan.

Before making any adjustments for labor, find out what tools and equipment are available for the job. Don't assume the equipment will always be in top operating condition. You may need to rent equipment when your owned equipment is not working or is in use elsewhere. In any event, only the good tools should be made available to the workers. Companies that are "tool poor" will always have low productivity.

Beware of the manufacturer's representative who claims to have a phenomenal labor-saving device. True, there are some, but most have limitations. There is always a better way to get the job done. But it doesn't make sense to spend two dollars to save one, unless of course you can save another dollar over and over again. As an estimator you should be cautious about predicting labor savings until you have seen your crews save those hours on several jobs.

You'll see new materials advertised as labor-saving alternatives to what you have been using. The trouble is that not all new materials comply with the code or meet job specs. Many inspection authorities are slow to accept new products and new application techniques. If a new product is installed and is later rejected by the inspector, you have a loss, not a gain. Reinstalling an approved material can delay other phases of the work.

New or Unusual Tasks

You're going to have new or unusual installation requirements that don't match any of the labor units in your file. Study the requirements and break down the installation into as many tasks as possible. Consider the size, weight and area of placement. Try to compare it to a similar task. Then make a reasonable estimate of the best possible labor unit.

If possible, follow up on the installation to verify your prediction. The manufacturer's agent or the sales representative may be able to give you an estimated installation time. If you weigh all factors, chances are that you'll be close to the right labor unit.

Breaking the job down into its components will usually reduce the chance of a large error in your estimate. Let's say that you're going to install a street light standard. All that is shown on the drawings is the symbol for the street light standard and a conduit run for the service. The job can be broken down as follows:

1. Luminaire—wattage, voltage, light distribution.

2. Lamp—type, wattage, voltage.

3. Photo cell control.

4. Pole—type, length, base, arm, handhole.

5. Anchor bolts—size, length, leveling nuts, shims.

6. Concrete pole foundation—size, depth, concrete mix strength.

7. Grout pole base to foundation.

8. Excavate for foundation.

9. Rebar cage, if specified.

10. Erection equipment.

11. Trenching, backfill, cleanup.

12. Conduit and fittings.

13. Wire for conduit and up to luminaire.

14. In-line fuse holder and fuse.

15. Ground rod—size, length.

16. Ground wire and conductors.

17. Splices.

18. Receiving and storage.

19. Placement at point of installation.

20. Testing.

Place each item on the pricing sheet so that the material price and labor units can be entered. If you find an item that you don't have a labor unit for, determine what a reasonable labor time would be. Enter that figure.

Here's an example. Consider an underground manhole or vault. You have these tasks:

1. Prefabricated manhole—size.

2. Unloading and placing.

3. Cover—size, traffic loading.

4. Collar—size, height.

5. Waterproofing.

6. Conduit entrance provisions.

7. Sump provisions.

8. Drainage.

9. Access ladder

10. Cable racks, hooks, insulators.

11. Ground rod and connectors.

12. Grounding of all metal hardware.

13. Cable tags.

14. Cable fireproofing.

15. Excavation.

16. Backfill and compaction.

17. Pavement removal.

18. Pavement restoration.

Most items identified on the plans can be broken down into many individual labor tasks so that the most accurate labor analysis can be made. Add all the labor elements together to get the composite labor time.

"S" is the symbol for a single pole wall-mounted switch. The job specifications usually identify the quality or grade, the color, the voltage, the ampere rating and possibly the manufacturer and catalog number.

But each "S" symbol is more than just a switch. The symbol probably assumes installation of:

1. The switch

2. Switch plate and trim

3. Outlet box

4. Outlet box cover or switch ring

5. Box hanger or blocking

When listing wiring devices on the pricing sheets, include only one item on each line. For a wall switch, for example, the single pole switches would be listed on one line. Each of the other parts should be listed under its own heading on your take-off sheet. The outlet box should be listed by type and size. The switch ring should be listed by type, depth and size. The boxes, rings, covers and backing must be listed separately because different locations require different accessories even though the same single pole switch is used.

There's another reason to list each part separately. Every part has its own labor requirement and price. Each switch is a very small part of your labor and material estimate. But your whole estimate is an accumulation of many very small parts. The more detailed and controlled the labor and material pricing, the more accurate the bid.

The example earlier in this book on labor units shows the following data for a single pole switch: S = Single pole switch at 0.25 hours plus 0.05 for the cover plate, or a total of 0.30 hours.

Plates should be listed separately when various colors and types are required.

Even labor units as small as 0.05 hours become significant costs when multiplied by the material quantity needed on a larger job.

Avoid using a lump sum labor quantity to adjust for the items you probably missed. Instead, try to identify every labor element. The only way to be sure you have included enough labor is to make an estimate of every labor item you can reasonably foresee.

You probably do want to include a lump sum allowance for small hardware items that can't be counted accurately and probably aren't worth counting even if you could. For these items, estimate your yearly cost and then divide by the number of jobs you expect to have in the coming year. The result is a lump sum to add to each estimate. Small hardware items include conduit fittings, machine screws, wood screws, lag screws, washers, rods, nuts, bolts, phase tape, plastic tape, masking tape, duct tape, spray paint, caulk, crayons, visqueen, pulling lube, cutting oil, rags, nails, rope, string, hangers, tie wire, toggle bolts, conduit seals, KO seals, pipe plugs, pipe caps, lugs, split bolt connectors, tags, tie wraps, nameplates, and reducing washers.

Summary
The only good estimate is a detailed estimate.

Remember that your estimate on every item you miss is always zero. That's a 100% miss.

Labor will be the hardest part of every estimate you make. Study the plans, know what your crews can handle, be familiar with your tool inventory, make reasonable estimates by dividing unfamiliar operations into familiar components, work carefully, and keep good records of actual performance. If you do all this, it's no accident that you've become a highly skilled electrical estimator.

Labor Unit Tables

The tables on this page and continuing to page 88 are labor units measured in tenths of an hour. See page 88 for the conversion table for minutes to decimal equivalents. Check the heading on each table for the type of material or equipment. Sub-headings indicate the installation variables.

The labor units can be used as your benchmark and adjusted from time to time to fit your company's experience, Some may be increased and others decreased after thorough study.

All of the benchmark labor units can be expanded on for a more accurate labor unit bank.

Labor to install galvanized rigid conduit, heavy wall

| | Per Hundred Feet | | | Each Fitting | | | |
	Embedded	Exposed	Concealed	Terminate	Elbow	Strap	Coupling
½"	3.50	4.00	3.75	.05	.05	.05	.05
¾"	3.75	4.50	4.00	.06	.06	.06	.06
1"	4.25	5.00	4.50	.08	.08	.08	.08
1¼"	5.00	6.00	5.25	.10	.10	.10	.10
1½"	6.00	8.00	6.50	.10	.10	.10	.10
2"	8.00	10.00	8.50	.15	.15	.10	.10
2½"	10.00	12.00	11.00	.15	.20	.10	.15
3"	12.00	14.00	13.00	.20	.20	.10	.15
3½"	14.00	16.00	15.00	.20	.20	.10	.15
4"	16.00	18.00	17.00	.25	.25	.15	.20
5"	20.00	24.00	22.00	.30	.30	.15	.20
6"	26.00	30.00	28.00	.40	.50	.20	.25

Labor units listed are average and must be factored for difficulty.
Multiple runs may save labor.

Labor to install galvanized rigid conduit, intermediate wall

	Per Hundred Feet			Each Fitting			
	Embedded	Exposed	Concealed	Terminate	Elbow	Strap	Coupling
½"	3.40	3.75	3.60	.05	.05	.05	.05
¾"	3.60	4.25	3.80	.06	.06	.06	.06
1"	4.00	4.50	4.20	.08	.08	.08	.08
1¼"	4.30	5.00	4.60	.10	.10	.10	.10
1½"	5.30	6.75	5.60	.10	.10	.10	.10
2"	7.00	8.00	7.50	.15	.15	.10	.10
2½"	9.00	10.00	9.50	.15	.20	.10	.15
3"	10.00	12.00	10.50	.20	.20	.10	.15
3½"	12.00	14.00	12.50	.20	.20	.10	.15
4"	14.00	16.00	14.50	.25	.25	.15	.20

Labor units listed are average and must be factored for difficulty.
Multiple runs may save labor.

Labor to install rigid aluminum conduit

	Per Hundred Feet			Each Fitting			
	Embedded	Exposed	Concealed	Terminate	Elbow	Strap	Coupling
½"	3.50	4.00	3.75	.05	---	.05	.05
¾"	3.75	4.50	4.00	.06	---	.06	.06
1"	4.25	4.75	4.50	.08	---	.08	.08
1¼"	4.75	5.50	5.00	.10	.10	.10	.10
1½"	5.75	7.00	6.00	.10	.10	.10	.10
2"	7.50	9.00	8.00	.15	.10	.10	.10
2½"	9.50	11.00	10.00	.15	.15	.10	.15
3"	11.50	13.00	12.00	.20	.20	.10	.15
3½"	12.50	14.00	13.00	.20	.20	.10	.15
4"	14.50	16.00	15.00	.25	.20	.10	.20
5"	17.00	19.00	18.00	.30	.25	.15	.20
6"	22.00	24.00	23.00	.40	.30	.20	.25

Multiple runs may save labor.

Labor to install flexible steel conduit, standard wall

	Per Hundred Feet		Each Fitting		
	Exposed	Concealed	Straight-Connector	Angle Connector	Coupling
½"	3.00	3.00	.05	.10	.05
¾"	3.25	3.25	.06	.10	.06
1"	3.75	3.50	.08	.10	.08
1¼"	4.00	3.75	.10	.15	---
1½"	4.50	4.00	.10	.15	---
2"	4.75	4.50	.15	.25	---
2½"	6.00	6.00	.15	.25	---
3"	7.00	7.00	.25	.50	---

Aluminum flexible conduit takes less labor, deduct 10%.

Labor to install electrical metallic tubing (EMT)

	Per Hundred Feet			Each Fitting			
	Embedded	Exposed	Concealed	Connector	Elbow	Strap	Coupling
½"	3.25	3.50	3.25	.05	---	.05	.05
¾"	3.50	4.00	3.75	.06	---	.06	.06
1"	4.00	4.50	4.25	.08	---	.08	.08
1¼"	4.50	5.00	4.75	.10	.10	.10	.10
1½"	5.50	6.00	5.75	.10	.10	.10	.10
2"	7.00	8.00	7.50	.15	.15	.10	.15
2½"	9.00	10.00	9.50	.20	.15	.10	.20
3"	10.00	12.00	11.00	.25	.20	.10	.25
4"	12.00	14.00	13.00	.25	.20	.10	.25

Multiple runs may save labor.

Labor to install polyvinyl chloride rigid conduit, Schedule 40

	Per Hundred Feet			Each Fitting			
	Embedded	Exposed	Concealed	Connector	Elbow	Strap	Coupling
½"	3.10	3.30	3.20	.05	.05	.05	.05
¾"	3.20	3.40	3.30	.05	.05	.05	.05
1"	3.30	3.50	3.40	.05	.05	.05	.05
1¼"	3.40	4.00	3.50	.10	.10	.10	.10
1½"	3.45	5.00	4.50	.10	.10	.10	.10
2"	3.50	6.00	5.00	.15	.15	.10	.10
2½"	3.65	8.00	6.50	.20	.15	.10	.10
3"	3.75	9.00	7.50	.20	.15	.10	.15
3½"	3.85	10.00	8.00	.25	.20	.10	.20
4"	4.00	11.00	8.50	.25	.20	.10	.20
5"	5.00	12.00	9.00	.30	.25	.15	.25
6"	8.00	15.00	12.00	.30	.25	.15	.25

Schedule 80 PVC is heavier, add 10% factor.
The fittings are used on either Schedule 40 or 80.
Multiple runs may save labor.

Labor to install power and communications and ABS duct

Per Hundred Feet		Each Fitting				
	Duct	End Bell	Adapter	Sweep	Base Spacer	Intermediate Spacer
2"	3.25	.10	.10	.15	.05	.05
3"	3.50	.15	.15	.20	.05	.05
3½"	3.60	.20	.20	.25	.05	.05
4"	3.80	.25	.25	.30	.05	.05
5"	4.00	.25	.25	.30	.05	.05
6"	5.00	.30	.30	.50	.05	.05

EB or DB take about the same labor.
Multiple runs may save labor.

Labor to install PVC coated rigid conduit, 40 mil

Per Hundred Feet		Each Fitting		
	Conduit	Elbow	Coupling	Cut/Thread
½"	4.50	.10	.10	.25
¾"	6.00	.10	.10	.25
1"	7.00	.15	.15	.30
1¼"	8.00	.20	.15	.40
1½"	10.00	.25	.15	.40
2"	12.00	.30	.20	.50
2½"	16.00	.40	.25	.75
3"	20.00	.50	.30	1.00
3½"	22.00	.60	.30	1.25
4"	24.00	.60	.30	1.30
5"	28.00	.75	.40	1.50

Multiple runs may save labor.
PVC coated rigid, 20 or 40 mil.
20 mil may save 10% labor.

Labor to install cable tray (trough type), 12' lengths

Item	12" Wide	18" Wide	24" Wide
Tray, 1 length	6.00	8.00	10.00
Elbow, 45⁰	1.00	1.50	1.75
Elbow, 90⁰	1.25	1.75	2.00
Tee	1.50	2.00	3.00
Cross	2.00	2.50	3.50
Dropout or end	.25	.30	.40
Adapter	1.00	1.50	2.00

Installation at 12' maximum in clear open areas.

Labor to install wireway (screw or hinged cover)

Size	Per Foot	Elbow	Tee	Flange	Coupling	End
4" x 4"	.25	.30	.50	.50	.10	.10
6" x 6"	.30	.40	.60	.60	.15	.15
8" x 8"	.50	.75	1.00	1.00	.15	.15
12" x 12"	.75	1.00	1.50	1.50	.25	.25

Wireway is indoor NEMA 1 type. Add for raintight applications.

Labor to install underfloor raceway

	Underfloor Duct			Trench Type	
	4"	8"	12"	18"	24"
Raceway 10'	.50	.75	2.50	4.00	6.00
Flat elbow 45⁰	.25	.30	1.00	1.25	1.50
Flat elbow 90⁰	.25	.30	1.00	1.25	2.00
Vertical elbow	.25	.30	---	---	---
Offset elbow	.30	.50	---	---	---
Tee	---	---	1.50	2.00	3.00
Junction box	1.00	1.00	---	---	---
Panel adapter	.50	1.00	2.00	3.00	4.00
Coupling, sleeve	.10	.10	---	---	---
Closure cap	.10	.10	---	---	---
Abandon cap	.25	.25	---	---	---
Conduit adapter	.10	.10	---	---	---
Marker assembly	.25	.25	---	---	---
Support, single	.10	.15	.25	.25	.25
Support, double	.15	.20	---	---	---
Service fitting	.50	.50	---	---	---

Installed on firm level areas up to 3rd floor. Add for other factors.

Labor to install wire, 1/conductor, solid or stranded, 600 volt

Per Thousand Feet

Size	Copper					Aluminum	
	TW	THW	THHN	XHHW	USE	THW	XHHW
14	5.50	6.00	5.50	6.50	6.50	---	---
12	6.50	7.00	6.50	7.50	7.50	---	---
10	7.50	8.00	7.50	8.50	8.50	---	---
8	9.50	10.00	9.50	11.00	11.00	---	---
6	11.50	12.00	11.00	13.00	13.00	8.00	9.00
4	13.50	14.00	13.00	15.00	15.00	10.00	11.00
2	15.50	16.00	15.00	17.00	17.00	12.00	13.00
1	---	18.00	16.00	19.00	19.00	14.00	16.00
1/0	---	20.00	18.00	21.00	21.00	16.00	17.00
2/0	---	22.00	20.00	23.00	23.00	18.00	19.00
3/0	---	24.00	22.00	25.00	25.00	20.00	21.00
4/0	---	26.00	24.00	27.00	27.00	22.00	23.00
250	---	28.00	26.00	29.00	29.00	24.00	25.00
300	---	30.00	28.00	31.00	31.00	26.00	27.00
350	---	32.00	30.00	33.00	33.00	28.00	29.00
400	---	34.00	32.00	35.00	35.00	30.00	31.00
500	---	36.00	34.00	37.00	37.00	32.00	33.00
600	---	38.00	---	39.00	39.00	33.00	34.00
750	---	40.00	---	42.00	42.00	34.00	35.00
1000	---	45.00	---	48.00	48.00	35.00	36.00

Labor units are based on 3 conductors per run.
Multiples thereon may save labor hours.

Labor to install high voltage wire, 1/conductor

Per Thousand Feet

Size	Copper, XLP or EPR		Aluminum, XLP or EPR	
	5 KV	15 KV	5 KV	15 KV
8	11.00	12.00	10.00	10.00
6	13.00	13.50	12.00	12.50
4	15.00	15.50	13.00	13.50
2	17.00	18.00	14.00	15.00
1	19.00	20.00	17.00	18.00
1/0	21.00	22.00	18.00	19.00
2/0	23.00	24.00	20.00	21.00
3/0	25.00	26.00	22.00	23.00
4/0	27.00	28.50	24.00	25.50
250	30.00	32.00	26.00	28.00
300	33.00	35.50	28.00	30.50
350	36.00	39.00	30.00	33.00
400	39.00	44.00	33.00	36.00
500	42.00	48.00	35.00	38.00
600	45.00	52.00	38.00	41.00
750	48.00	56.00	40.00	44.00
1000	55.00	65.00	45.00	50.00

Labor units are based on 3 conductors per run.
Multiples thereon may save labor.

Labor for splicing and terminations, 1/conductor

Size	Insulated Connector	Split Bolt	Compress Sleeve	Epoxy Kit	Thru 5 KV	Wye 5 KV	Thru 15 KV	Wye 15 KV
14	.03	---	.03	---	---	---	---	---
12	.03	---	.03	---	---	---	---	---
10	.05	.10	.05	---	---	---	---	---
8	---	.10	.10	.30	2.50	3.00	2.50	3.00
6	---	.15	.15	.35	2.75	3.25	3.00	3.50
4	---	.15	.15	.35	2.75	3.25	3.00	3.50
2	---	.20	.20	.50	3.00	3.50	3.25	3.75
1	---	.25	.25	.60	3.00	3.50	3.25	3.75
1/0	---	.30	.30	.75	3.25	3.75	3.50	4.00
2/0	---	.40	.40	1.00	3.25	3.75	3.50	4.00
3/0	---	.50	.50	1.00	3.50	4.00	3.75	4.25
4/0	---	.60	.60	1.25	3.50	4.00	3.75	4.25
250	---	.75	.75	1.50	3.75	4.25	4.00	4.50
300	---	1.00	1.00	2.00	4.00	4.50	4.25	4.75
350	---	1.25	1.25	2.50	4.00	4.50	4.25	4.75
400	---	1.25	1.25	2.50	4.00	4.50	4.25	4.75
500	---	1.50	1.50	3.00	4.50	5.00	4.75	5.50
600	---	1.50	1.50	3.00	4.50	5.00	4.75	5.50
750	---	2.00	2.00	4.00	5.00	6.00	5.50	6.50

The splices and termination on over 600V circuits are generally made by qualified splicers. The splicer's pay rate is about 10% higher than an electrician's rate.

Labor to install floor box (adjustable)

Type	Box	Cover	Device	Mound	Plug
Steel	.50	.30	.20	.30	.10
Round cast	.60	.30	.20	.30	.10
Square cast	.75	.30	.20	.30	.10
Double cast	1.00	.50	.40	---	.20
Triple cast	1.50	.75	.50	---	.30

Installed on firm level areas. Add if in grade beams, etc.

Labor to install junction boxes

Size	Screwdriver	Flush	Raintight	JIC	Cabinet
6" x 6" x 4"	.40	.50	.50	.50	---
8" x 8" x 4"	.50	.60	.60	.60	---
8" x 8" x 6"	.60	.70	.70	.70	---
10" x 10" x 4"	.65	.75	.75	.75	---
10" x 10" x 6"	.65	.75	.75	.75	---
12" x 12" x 4"	.70	.80	.80	.80	1.00
12" x 12" x 6"	.75	1.00	1.00	1.00	1.25
12" x 12" x 8"	.80	1.00	1.00	1.25	1.50
18" x 18" x 4"	1.25	1.50	1.50	2.00	2.50

All are code gauge steel.

Labor to install bus duct (plug-in), 3 phase, 48 OV

Item	60	100	225	400	600	800	1000	2000	3000
10' copper	.80	1.00	2.00	2.50	3.00	3.50	3.75	5.00	6.00
10' aluminum	---	.75	1.00	1.50	2.00	2.50	3.00	4.00	5.00
Flat elbow	.25	.40	.75	1.25	1.50	3.00	4.00	6.00	7.00
Vertical elbow	.25	.40	.75	1.25	1.50	3.00	4.00	6.00	7.00
Tee	.30	.50	1.00	1.50	2.00	2.50	3.50	6.50	8.00
Panel connector	.25	.50	1.00	1.00	1.50	2.00	2.50	4.00	5.00
Wall flange	---	.50	1.00	1.00	1.50	2.00	2.00	3.00	4.00
Cable tap box	.30	.75	1.50	2.00	2.00	3.00	4.00	4.50	5.00
End closure	.10	.15	.20	.50	.50	.50	1.00	1.00	1.00
Entrance cover	---	.10	.10	.15	.15	.15	.20	.20	.20
Hanger assembly	.15	.25	.25	.30	.50	.50	1.00	1.50	2.00
100A frame Sw.	---	.50	.60	.60	.60	.60	.60	.60	.60
200A frame Sw.	---	---	1.00	1.00	1.00	1.00	1.00	1.00	1.00
400A frame Sw.	---	---	---	1.50	1.50	1.50	1.50	1.50	1.50
600A frame Sw.	---	---	---	---	2.00	2.00	2.00	2.00	2.00

Installed up to 12' above floor in open areas.

Labor to install outlet boxes and covers

	Each Fitting						
	Box		Cover				
Box	Depth 1½"	Depth 2½"	Plaster Ring	1 Gang Switch	2 Gang Switch	Blank Plate	Industrial Raised
Handy	.15	.15	.05	.05	---	.05	.10
3/0	.15	.15	.05	---	---	.05	---
4/0	.20	.20	.05	---	---	.05	---
4/S	.25	.25	.05	.05	.06	.05	.10
5/S	.30	.30	.06	.06	.08	.06	.15
3/G	.35	.40	---	---	---	---	---
4/G	.40	.50	---	---	---	---	---
5/G	.50	.60	---	---	---	---	---
6/G	.50	.60	---	---	---	---	---

Outlet boxes are pressed steel. Non-metallic boxes may be installed at a savings of 10%.

Labor to install wiring devices and plates

Item	Device	Plate
Single-pole switch	.25	.05
Double-pole switch	.30	.05
Three-way switch	.30	.05
Four-way switch	.50	.05
Weather-proof switch	.25	.10
Door switch	.50	.05
Duplex receptacle	.25	.05
Single receptacle	.20	.05
Weather-proof receptacle	.25	.10

Labor unit includes the wiring device, connecting the wires, installing in box and attaching the cover.

Labor to install motor hookup and control equipment

H.P.	Hookup	Disconnecting Switch	Circuit Breaker	Magnetic Starter	Manual Starter	Lockout Stop Switch
To ½	.50	.30	.50	.75	.50	.30
¾	.50	.30	.50	.75	.50	.30
1	.60	.30	.50	.75	.50	.30
1½	.75	.30	.50	.75	.50	.30
2	.75	.30	.50	.75	.50	.30
3	.75	.30	.50	.75	.50	.30
5	1.00	.50	.60	.75	.50	.30
7½	1.00	.75	.60	1.00	----	.30
10	1.25	.75	.60	1.00	----	.30
15	1.50	.75	.75	1.50	----	.30
20	2.00	1.00	1.00	1.50	----	.30
25	2.50	1.00	1.25	1.50	----	.30
30	3.00	1.50	1.50	2.50	----	.30
40	4.00	2.00	1.50	2.50	----	.30
50	5.00	2.00	1.50	2.50	----	.30
60	6.00	3.00	2.00	4.00	----	.30
75	7.00	4.00	2.00	4.00	----	.30
100	8.00	4.00	2.50	4.00	----	.30
125	8.00	5.00	3.00	6.00	----	.30
150	10.00	5.00	3.50	6.00	----	.30

Each motor hookup includes the flexible connection, make-up of wires, check rotation and proper operation.

Hand-off-auto switch	.40	Push button (stop-start)	.40
Selector switch	.30	Pilot light	.25

Labor to install transformer (dry type), below 600V, 3 phase

Size	Wall Mounted	Floor Mounted	Pad Mounted
3 KVA	1.00	---	---
5 KVA	1.50	---	---
7½ KVA	2.00	---	---
10 KVA	2.00	1.50	---
15 KVA	2.50	1.75	---
25 KVA	3.00	2.00	2.50
37½ KVA	3.50	2.50	3.00
45 KVA	3.50	2.50	3.00
75 KVA	4.00	3.00	4.00
100 KVA	4.00	3.00	4.00
112½ KVA	4.50	4.00	5.00
150 KVA	5.00	4.00	5.00
225 KVA	---	6.00	7.00
300 KVA	---	6.50	8.00
500 KVA	---	6.50	8.00
750 KVA	---	8.00	10.00
1000 KVA	---	12.00	16.00

Add for difficult locations, such as basements, roofs, ceiling spaces, etc.

Labor to install panelboard, loadcenter

Ampere	Circuit Breakers Below 600 Volt		
	1 Pole	2 Pole	3 Pole
15	.30	.40	.50
20	.30	.40	.50
30	.40	.50	.60
40	.50	.60	.75
50	.50	.60	.75
60	.50	.60	.75
70	.60	.75	1.00
100	.75	1.00	1.50
125	---	1.00	1.50
150	---	1.50	2.00
175	---	1.50	2.00
200	---	2.00	2.50
225	---	2.50	3.00
250	---	2.50	3.00
300	---	3.00	3.50
350	---	3.00	3.50
400	---	3.50	4.00
450	---	3.50	4.00
500	---	4.00	5.00

Add additional labor for installing circuit breakers in panelboards or loadcenters.

Enclosures	Nema 1	Nema 3R
To 100A	.30	.40
225A	1.00	1.50
400A	2.00	3.00

Enclosures are for circuit breakers only.

Labor to install distribution equipment

Panel, Can Only	Surface	Flush
12 circuit can	.50	.70
24 circuit can	.75	1.00
30 circuit can	1.00	1.50
42 circuit can	1.50	2.00

The interior and trim portions are not included above.

Labor to install metering equipment

Number of Units	Indoor	Raintight
1 service	.75	.75
2 services	1.00	1.00
3 services	1.50	1.50
4 services	2.00	2.00
5 services	4.00	4.50
6 services	4.50	5.00
7 services	5.00	5.50
8 services	5.50	6.00
10 services	8.00	---

Includes total installation and connections.

Labor to install surface metal raceway

Item	½"	¾"	1"	1¼"
10' length	.30	.40	.50	.60
Tee	.10	.10	.15	.20
Elbow	.10	.10	.15	.20
Adapter, conduit	.15	.15	.20	.20
Adapter, box	.10	.10	.15	.15
Adapter, plate	.15	.15	.15	.20
Coupling	.05	.05	.05	.05
End fitting	.05	.05	.05	.05
Outlet box	.20	.20	.20	.20
Switch box	.20	.20	.20	.20
Extension box	.20	.20	.20	.20
Distribution box	.25	.25	.25	.25

Similar to Wiremold Products. Installation on clean flat surfaces not over 8 feet above finish floor.

Labor to install incandescent lighting fixtures

Item	Open	Enclosed	Pendant	Recessed	Industrial	Flood
100W 1 lamp	.30	.40	.40	.50	.40	.40
100W 2 lamp	---	.50	.50	.60	---	.50
150W 1 lamp	.30	.40	.50	.50	.50	.50
150W 2 lamp	---	.50	.50	.60	---	.60
300W 1 lamp	.40	.70	.60	.70	.60	.75
300W 2 lamp	---	---	---	---	---	.80
500W 1 lamp	.50	1.00	1.00	1.10	1.00	1.25

Installed in areas below 8 feet above finish floor.

Labor to install fluorescent light fixtures

Item	Open	Enclosed	Pendant	Recessed	Industrial
24" 1 lamp	.40	.50	.60	.60	.50
24" 2 lamp	.45	.55	.65	.65	.55
36" 1 lamp	.50	.60	.65	.70	.60
36" 2 lamp	.55	.65	.70	.75	.65
48" 1 lamp	.50	.60	.65	.70	.60
48" 2 lamp	.55	.65	.70	.75	.65
72" 1 lamp	.60	.70	.70	.80	.70
72" 2 lamp	.65	.75	.75	1.00	.75
96" 1 lamp	.80	1.00	1.25	1.50	1.00
96" 2 lamp	1.00	1.25	1.50	1.75	1.25

Labor to install lamps

Incandescent	.05
Fluorescent	.05
Special	.10

Add additional labor for relamping jobs.

Labor to install signal devices

Single Devices	Unit Price	Single Devices	Unit Price
Surface clock, 12"	.30	Fire alarm horn, single projection	.30
Flush clock, 12"	.50	Fire alarm horn, double projection	.50
Surface clock, 15"	.30	Fire alarm control panel	5.00
Flush clock, 15"	.50	Fire alarm pedestal and light	2.50
Skeleton clock, 24"	1.50	Fire alarm supervisory panel	2.50
Program buzzer	.20	Fire alarm isolating transformer	1.00
Program bell, 6"	.25	Fire alarm annunciator, 12 lamp	4.00
Program bell, 10"	.30	Fire alarm master coded station	2.50
Call chime	.25	Fire alarm evacuation siren	.50
Master clock	6.00	Fire alarm smoke detector	.40
Impulse generator	4.00	Fire alarm door holder	1.00
Fire alarm manual station	.40	Fire alarm air shutdown switch	1.50
Fire alarm bell, 4"	.25	Sprinkler alarm switch	1.50
Fire alarm bell, 6"	.25	Sprinkler pressure switch	1.50

Each device installation includes connecting the wire and securing the device.

Conversion table for minutes to decimal equivalents

Minutes	Decimals	Minutes	Decimals	Minutes	Decimals
1	.017	21	.350	41	.683
2	.033	22	.367	42	.700
3	.050	23	.383	43	.717
4	.067	24	.400	44	.733
5	.083	25	.417	45	.750
6	.100	26	.433	46	.767
7	.117	27	.450	47	.783
8	.133	28	.467	48	.800
9	.150	29	.483	49	.817
10	.167	30	.500	50	.833
11	.183	31	.517	51	.850
12	.200	32	.533	52	.867
13	.217	33	.550	53	.883
14	.233	34	.567	54	.900
15	.250	35	.583	55	.917
16	.267	36	.600	56	.933
17	.283	37	.617	57	.950
18	.300	38	.633	58	.967
19	.317	39	.650	59	.983
20	.333	40	.667	60	1 hour

All labor units are measured in manhours and given in hours and decimals of an hour.

Chapter 11

Commercial Take-off

This chapter takes you step by step through the take-off of a 6,000 square foot commercial building. Review the plans and specifications on the following pages. Then try to take off the materials and add prices and labor estimates. A complete estimate is included in this chapter so you can verify your answers against the actual bid on the project.

The 6,000 square foot building is to be constructed in an industrial park within the city limits. Construction is to start the first week in May and must be completed within 180 calendar days. Liquidated damages are set at $500.00 for each day of delay.

Owner will arrange for and will pay all utility company fees.

Architect will assign a resident inspector during construction. Permits and inspection from all regulating authorities are required. Signed off permits are a condition of completion.

General Contractor will supervise the construction of the building and is responsible for all construction utilities.

Building exterior walls are unit masonry. Interior walls are metal studs with gypsum board lath. The roof is built-up insulation hot mopped and gravel. Office area ceilings are suspended exposed grid system with 2 foot x 4 foot

acoustical panels

Site is level on a paved street. The soil is medium firm, no large rocks. Access is clear.

Mechanical Contractor is responsible for all controls not shown on the electrical drawings.

Section 16A Electrical
General. The applicable provisions of the General Conditions and the Special or Supplementary Conditions shall govern the work of this section the same as though written herein in full.

Scope of Work
A. The Contractor agrees that his submitted price for the work hereunder includes sufficient money allowance to make his work complete and operable. He agrees that inadvertent discrepancies or omissions, failure to show details, or repeat notes or figures on each drawing will not be the cause for additional charges or claims.

B. The electrical drawings shall be considered part of these specifications, and any work or materials shown on the drawings and not mentioned in the specifications, or vise versa, shall be executed as if specifically mentioned in both.

C. Minor changes to accommodate the installation of the work with other work of the Contract, shall be made by the Contractor when ordered by the Architect, without additional cost to the Contract.

D. The work under this section of the specifications shall include the furnishing of all equipment, apparatus, tools, labor and material required to complete the electrical installation for all that is indicated or specified herein.

E. Regulations, Codes, Permits. All work and materials included in this section shall be in accordance with the latest rules and regulations of local and state authorities, State Fire Marshal, OSHA and any other legally constituted Public authority having jurisdiction. Nothing in the plan or specifications shall be construed as permitting work that is not in conformance with applicable codes. The Contractor shall apply and pay for all required permits, arrange and pay for inspections, and deliver certificates of all such inspections to the Architect.

F. This specification and the drawings cover the complete electrical system and all related work including, but not limited to, the following:

1. The furnishing and installing of the electrical service to the building (excluding conductors), panelboards, lighting fixtures and lamps, conduits, outlets, wiring and wiring devices as required for a complete installation.

2. The supplying, installation, and complete connection of all disconnect switches and all magnetic starters that are shown on the drawings, and the installation and complete connection of all motor starters where shown on the drawings.

3. The complete connection of all motors.

4. Certain conduits and wiring for mechanical system control, where such is shown on the electrical drawings. All other control wiring for the mechanical system shall be furnished and installed under the Mechanical Sections of these specifications.

5. Complete provisions for the Telephone Company, including the service conduit.

6. The supplying and installation of nameplates for all panels, disconnect switches, motor starters, special outlets and terminal cabinets.

G. Substitution of Materials. Notwithstanding any reference in the specifications to any article, device, product, material, fixture, form, or type of construction by name, make, or catalog number, such reference shall be interpreted as establishing a standard of quality and shall not be construed as limiting competition; and the Contractor in such cases, may at his option use any article, device, product, material, fixture, form, or type of construction which in the judgment of the Architect expressed in writing, is equal to that specified.

H. Examination of Jobsite. Before commencing work the Contractor shall inspect the site. By the act of submitting a bid price for the work, the Contractor shall be deemed to have verified existing conditions and that he is familiar with and accepts all conditions at the jobsite.

Scaled and computed dimensions are approximate and are given for estimating purposes only. Before proceeding with the work, the Contractor shall carefully check and verify all dimensions and sizes of equipment and shall assume all responsibility for the fitting of his materials and equipment to other parts of the equipment and to the structure. He shall verify that the equipment submitted will fit the allotted spaces. If spaces are not adequate for equipment or materials, the Contractor shall pay for all alterations required to permit installation. Where work requires connection to NIC equipment and other equipment that is furnished and set by others, the Contractor shall obtain exact rough-in dimensions from those who are furnishing the equipment and shall install connections in a neat and workmanlike manner.

I. Locations. The locations of conduits, outlets, apparatus and equipment indicated on the drawings are approximate only and shall be changed, as approved by the architect, to meet the architectural and structural conditions as required. Locations indicated on the drawings, however, shall be adhered to as closely as possible.

All conduit and equipment shall be installed such as to avoid obstructions, preserving head room and keeping openings and passageways clear. Lighting fixtures, switches, convenience outlets, etc., shall be located within finished rooms, as indicated on the architectural drawings. Where these drawings do not indicate exact locations, such locations shall be obtained from the architect. Where equipment is installed without instruction and must be moved, it shall be moved without additional cost to the contract.

J. Drawings and Specifications are intended to complement each other. All miscellaneous items of work and materials necessary for the completion of the installation of the systems shall be provided whether or not mentioned in the specifications or shown on the drawings. Where a conflict exists between the requirements of the specifications and the drawings, the architect shall determine the intended requirements.

The architect shall interpret the drawings and the specifications, and his decision as to the true intent and meaning thereof and the quality, quantity and sufficiency of the materials and workmanship furnished thereunder shall be accepted as final and conclusive.

K. Excavations. All excavating, trench work and backfilling that is required for the installation of the work shall be performed under this section and in accordance with the applicable portions of the section of these specifications and plans on the subject of excavation, trenching and backfill.

L. Concrete Work. All concrete required for the installation of the electrical systems shall be performed under this section and in accordance with the applicable portions of the section of these specifications and drawings on the subject of concrete work.

M. Materials and Approvals. All materials furnished under this specification shall be new, in accordance with the specifications of the Institute of Electrical and Electronics Engineers, National Electrical Manufacturer's Association, National Fire Protection Association and the National Electrical Code, and shall bear the Underwriter's Laboratories label where such service is regularly available.

All equipment used for the same purpose shall be of the same make.

Wherever on the drawings, or in these specifications, materials are identified by the name of one manufacturer, it is intended that approved equivalent materials of other manufacturers are acceptable. However, where three or more manufacturers are mentioned, then the equipment furnished must be the product of one of the manufacturers listed.

Within 35 days after award of the contract, the contractor shall deliver to the architect a complete list of all materials, equipment, apparatus and lighting fixtures which he proposes to use. The list shall include size, type, name of manufacturer, catalog number and such other information required to identify the item.

No material, equipment, apparatus or lighting fixture shall be purchased or installed unless and until such materials, etc. have been approved by the architect. Any and all materials installed in violation of this provision shall, when so directed by the architect, be removed and replaced with materials acceptable to the architect at the contractor's expense.

N. Protection of Finish. The contractor shall provide adequate means for and shall fully protect all finished parts of materials and equipment against damage from any cause during the progress of the work and until acceptance by the architect.

Materials and equipment, both in storage and during construction shall be covered so that finished surfaces will not be damaged, marred or splattered with plaster or paint, and moving parts shall be kept clean and dry.

Damaged materials or equipment including face plates or panels and switchboard sections shall be replaced or refinished by the manufacturer at no additional cost to the contract.

O. Cleaning Equipment and Premises. All parts of the materials and equipment shall be thoroughly cleaned. Exposed parts shall be cleaned, all grease and oil spots shall be removed.

Exposed unfinished surfaces that are to be painted at the jobsite shall be cleaned and all rust, dust, oil, grease and other marks shall be removed prior to painting.

During the progress of the work the contractor shall clean up after the installers and shall leave the premises and all portions of the building in which he is working free from his debris.

P. Verbal Agreements. No verbal agreement shall affect or modify any of the terms, clauses, instructions, notes or details that are contained in the documents comprising the contract. All such agreements shall be in writing before becoming a part of the contract.

Q. Cooperation with Other Trades. The contractor shall cooperate with other trades on the project for the proper execution of the work. Refer to the architectural, structural and mechanical sections and drawings for construction details, hookups and connections.

R. Guarantee. In addition to the guarantees required in the general provisions, all materials and equipment provided and installed under this section shall be guaranteed for a period of one (1) year from the date of acceptance of the work by the architect. Should any trouble develop during this period, due to defective materials, faulty equipment or poor workmanship, the contractor shall furnish all necessary labor, materials or equipment to correct the trouble, without additional cost to the contract.

S. Tests. All systems shall test free from short circuits, open circuits, wrong connections and grounds, free from mechanical and electrical defects and shall show an insulation resistance between phase conductors and ground not less than that required by the National Electrical Code. All circuits shall be tested for proper neutral connections. Furnish all necessary instruments and equipment required for making tests, and immediately correct any defective work.

Upon completion of this work and adjustment of all equipment, all systems shall be tested in the presence of the architect to demonstrate that all equipment furnished and installed or connected under the provisions of this contract function electrically in the manner specified.

T. Record and As-built Drawings. The contractor shall provide and maintain in good order in his field office a complete set of blue line prints or all electrical drawings which form a part of the contract. In event any work is not installed as indicated in the drawings, such work shall be carefully and correctly drawn on these as-built prints.

As-built drawings shall be provided in accordance with the general provisions of the contract.

U. Submittals. The following information shall be furnished as submittal data to the architect for approval before installation.

  1. Catalog Sheets
  a. Lighting fixture, (each type)
  b. Disconnect switches, (each type)
  c. Motor control equipment, (each type)
  2. Shop Drawings
  a. Main switchboard
  b. Panelboards, (each one)
  3. Material Listing
  a. All material specified in this section.

  b. All material specified in the contract drawings

Materials and Equipment

1. Conduit shall be delivered to the site in standard lengths and each length shall bear the manufacturer's trademark or stamp and the Underwriter's label of approval.

2. Rigid conduit shall be hot dipped galvanized or sheradized. Couplings, locknuts, bushings, etc., shall be hot dipped galvanized or sheradized.

3. Electrical metallic tubing shall be galvanized or sheradized. Couplings and connectors shall be the compression type or water tight. Do not use "indent" or "set screw" type.

4. Flexible conduit shall be galvanized or sheradized. Connectors shall be "squeeze tight" or "jake" type.

5. Oil tight flexible conduit shall be galvanized or sheradized with an outer neoprene jacket of approved type complete with a grounding conductor. Connectors shall be the type suitable for oil tight flexible conduit.

6. Non-metallic conduit shall be polyvinyl chloride (PVC), Schedule 40, U.L. approved. Couplings, adapters, elbows and all other fittings shall be U.L. approved and of the same type of material.

7. Wire shall be copper insulated (THW, THHN or XHHW) delivered to the jobsite in unbroken packages, plainly marked or tagged as follows:

  a. Underwriter's label.
  b. Size, kind and insulation type.
  c. Name of manufacturer and trade name of wire.
  d. Month and year when manufactured. The date shall not exceed one (1) year prior to the delivery date. Minimum size shall be No. 12 AWG, unless otherwise specified.

8. Lighting fixtures shall be as shown on the lighting fixture schedule in the electrical drawings.

All fixtures shall be new and in their original package.

All fixtures shall be wired from the outlet to the socket with No. 14 AWG Underwriter's type "AF" fixture wire, except that conductors in wiring channels of fixtures mounted in continuous rows shall be 600 volt "RHH" and of the same size as the circuit conductors.

All fixtures shall be equipped with lamps of the size and type as indicated or specified.

Lamps shall be General Electric, Westinghouse, or Sylvania. Fluorescent lamps shall be warm white lamps.

All fluorescent fixture ballasts shall be H.P.F., C.B.M. approved and E.T.L. certified Class "P" premium quality.

9. Outlet boxes shall be of proper code size for the number of wires or circuits passing through or terminating therein.

Flush outlet boxes shall be pressed steel, knockout type, hot dipped galvanized or sheradized. Use approved factory made knockout seals in all boxes where knockouts are not intact. Use outlet boxes as pull boxes wherever possible. Outlets required for flush mounting shall be furnished with a plaster ring. Install single or multi-gang boxes as required for the quantity of conductors and/or devices to be mounted. Light outlet boxes shall be equipped with plaster rings.

Surface mounted outlet boxes shall be cast metal boxes equipped series "FS" or "FD" condulets or shall be one piece pressed steel with raised surface covers.

Flush mounted boxes shall be set so that the plaster ring is even with the finished surface. Plaster rings shall be properly sized for each wall covering thickness.

10. Floor boxes shall be watertight, adjustable type, arranged so that the top may be varied from the plane of its base. Cover plates shall be heavy brass with permanent ring or flange and gasket. Boxes shall be National Series 800 or equal. Flush power outlets shall be National No. 8255R2. Install approved carpet flanges where required.

11. Single pole switches shall be specification grade equal to Leviton 1221-I, Hubbell 1221-I or Sierra 5021.

12. Convenience outlets shall be specification grade, flush mounted, duplex with parallel blades and "U" ground, rated for 15A, 125 volts, designed for side or back wiring, ivory, equal to Leviton 5242-I, Hubbell 5242-I or P and S 5242-I.

Weatherproof convenience outlets shall have an approved vertical weatherproof cover made of an insulating material.

Ground fault interrupters shall be duplex equal to Slater No. SIR-15-F-IV.

Special outlets shall be as indicated on the drawings. 30A shall be equal to Hubbell No. 2610. 50A shall be equal to Hubbell No. 9360. Special outlets shall have approved ivory cover plates that properly fit the receptacle.

13. Cover plates shall be smooth ivory equal to Leviton No. Series 87000, except surface mounted boxes shall have galvanized plates.

14. Plugmold shall be equal to Wiremold No. 20G12 for single circuits and No. 20GA12 for two circuits. Provide Wiremold fittings, joiners, end covers, adapters, straps and all necessary fittings for a complete installation.

15. Service distribution main switchboard shall be floor standing with an underground pull section, metering, C.T. and main circuit breaker section and a distribution section.

Submit shop drawings to the utility company for approval before fabrication and comply with all requirements.

Circuit breakers for the distribution switchboard shall be of the molded case type. Size as indicated on the drawings. The trip setting shall be in accordance with electrical codes. The breakers shall be operated by a toggle mechanism and shall be of the quick-make, quick break type and entirely trip free. The circuit breakers shall be provided with inverse time thermal overload and with instantaneous short circuit protection. All poles of the circuit breaker shall trip simultaneously on overload or short circuit on any pole. The operating handle shall clearly indicate the position of the breaker. Means shall be provided to lock each circuit breaker handle in the off position by means of several padlocks. Keyed handles will not be permitted.

16. Panelboards shall be of the automatic circuit breaker type, quick-make and quick break, trip free, inverse time characteristic thermal and magnetic tripping elements. Circuit breakers shall be of the bolted-on type, Square D type NQOB or approved equal.

Multi-pole circuit breakers shall be common trip, single handle.

Locking devices shall be of the lock-on lock-off type and located as indicated in the drawings.

Trims shall have doors equipped with flush type combination locks and catch, two-milled type keys supplied with each lock. All locks shall be keyed alike. A plastic covered directory card containing all numbered circuits in the panelboard shall be mounted inside the panel trim door in a metal frame. The contractor shall neatly type in each circuit that is connected in the panelboard.

Panelboards shall be shop primed and finished to match surrounding finishes after installation. Color shall be as approved by the architect. Provide nameplates with identification of panel and install on the face of the panel,

17. Terminal cabinets shall be surface mounted with trim cover door equipped with flush type combination lock and catch. A 3/4-inch plywood backboard shall be installed inside the cabinet. The terminal cabinet shall be 12 inches wide, 18 inches high and 4 inches deep. Finish shall be the same as for panelboards.

18. Telephone service conduits, backboard and terminal cabinets shall be in accordance with the requirements of the telephone company. The conduit systems shall be inspected and approved by the telephone company before being covered or concealed.

19. Time switches shall be surface mounted, NEMA 1, 120V, 40A, 2 pole, astronomical dial, day omitting, reserve power and equal to Tork No. 7202ZL.

20. Safety switches shall be as indicated in the drawings and shall be heavy duty, motor rated, fused or nonfused, NEMA 1 or NEMA 3R and shall be equal to Square D and NEMA KSI-1975 for type HD. Fusible switches shall be complete with fuses as required by the equipment manufacturer.

21. Push button stations shall be equal to Hubbell No. 1251 MC. Push button switches mounted on the exterior of the building shall have a cover equal to Hubbell No. 1750.

22. Grounding shall comply with the applicable requirements of the National Electrical Code. The main switchboard shall be grounded by means of a "Ufer" grounding electrode. The Ufer ground shall be No. 3/0 B.C. An additional ground shall be connected to the serving end of the water meter by means of approved connectors. The water pipe ground shall be No. 3/0 B.C.

23. Ground tests shall be made in the presence of the resident inspector. Resistance shall be measured with an instrument developing 500 volts and designed for accurately indicating the resistance. Test reports shall be prepared by the contractor and submitted to the architect. The tests shall include the measured resistance between each phase to phase and each phase to ground. All conductor insulation values shall be in accordance with the National Electrical Code.

The test report shall indicate the feeder circuits that are tested and each phase shall be identified as Phase "A," Phase "B" and Phase "C."

All equipment and devices shall be demonstrated to operate in accordance with the specifications requirements.

All test equipment shall be furnished by the contractor.

Installation

1. The minimum sizes of conduits for the various circuits and feeders shall be as indicated on the drawings, or code size for the number and size of conductors shown, unless a larger size is indicated, in which case such larger size shall be used. Open ends shall be capped with approved manufactured conduit seals as soon as installed and kept capped until ready to pull in conductors. Bushings shall be installed on all conduit ends before conductors are pulled. Running thread connections shall not be permitted. Use all approved manufactured conduit fittings.

2. Standard weight galvanized or sheradized rigid steel conduit shall be used where exposed to the weather, placed in masonry or concrete construction or where subject to mechanical injury.

3. Electrical metallic tubing (EMT) may be used in interior dry non-masonry walls and ceilings. EMT may be used in exposed interior locations where not subject to mechanical injury.

4. Polyvinyl chloride conduit (PVC) may be installed directly under the building grade slab and in other underground locations. A code size bonding conductor shall be installed with the other circuit conductors. The bonding conductor shall be colored green insulation or shall have green plastic tape bands to indicate that the conductor is the bonding conductor. All conduits that are installed underground outside the building area shall be buried to a depth of at least 24 inches. Utility conduits shall be buried to a depth acceptable to the utility company.

5. Underground conduits shall be encased in a concrete envelope of at least three inches from each side of the conduit. When more than one conduit is encased in the same envelope, they shall be separated by 1½ inches. Provide adequate chairs or spacers to support the conduit in the correct envelope location. All encasement concrete shall be 3,000 psi and the rock aggregate shall not exceed 1 inch in size.

6. Flexible metallic conduit shall be used to make connections to motors and other vibrating equipment. Flexible conduit shall be

used to connect lay-in fixtures to the junction boxes. Flexible conduit used in damp or exterior locations shall be of the oil tight type with a neoprene jacket.

7. Empty conduits shall have a nylon pull string installed from end to end with at least 24 inches of slack at each end. The tensile strength of the string shall be at least 200 lbs.

8. Conduit shall be supported with formed galvanized clamps, hangers, brackets, pipe straps or trapezes at intervals not exceeding eight (8) feet and in all cases with support not more than eighteen (18) inches from any outlet and at any point where it changes in direction.

Conduits over metal channel, lath and plaster ceilings shall be tied to the furring channels with No. 16 gauge galvanized wire ties spaced not more than five (5) feet apart. Conduits over suspended exposed grid system ceilings shall be supported with approved conduit supporting devices attached to the suspended ceiling wires. Conduits over 1 inch in size shall not be supported from the ceiling grid wires.

9. Whenever exposed or concealed conduits cross or intersect expansion joints in the building, suitable expansion fittings shall be provided.

Exposed or concealed conduits, fittings or boxes shall not be fastened to or come into contact with any water piping, heating ducts or other similar work, unless shown on the drawings or directed by the architect. It shall not be run above or adjacent to such piping or ducts and shall be supported independently. A minimum clearance of two (2) inches shall be maintained and in the case of uncovered hot water piping the minimum clearance shall be six (6) inches.

10. Where two or more conduits are exposed in a parallel run, an approved trapeze hanger shall be used.

11. Metal sleeves shall be provided where conduits pass through masonry or concrete walls. Sleeves shall be No. 12 gauge galvanized steel and shall be not more than 1/2 inch greater in diameter than the outside diameter of the conduit. Sleeves above and below grade shall be caulked watertight with "Duxseal."

12. All conduits passing through the roof shall be roof flashed and counter-flashed and made watertight.

13. The electrical contractor shall perform the work of cutting and patching of pavements, curbs, sidewalks and gutters wherever necessary for the installation of conduits. Damages to existing surfaces shall be repaired to match existing.

14. All wiring shall be installed in metallic and non-metallic raceways in accordance with the National Electrical Code, and latest adopted revisions.

Pull-in compound or lubricant shall be Ideal 77 Yellow or equal. Mechanical wire puller may be used where directed.

15. Splices in wires shall join the conductors securely together both mechanically and electrically. Wire size No, 10 and smaller shall be spliced with Minnesota Mining Company insulated "Scotch-lok" connectors, or equal. Wire sizes No. 8 and larger shall be joined together by means of Burndy split-bolt connectors and shall be neatly covered with insulating tapes or material to equal the wire insulation value. All splices shall be made in accordance with the National Electrical Code.

16. Lighting fixture shall be installed plumb, straight and level where shown on the drawings or where directed by the architect.

All lighting fixtures shall be complete with all fittings and parts necessary to completely and properly install as required.

The electrical contractor shall provide and install all blocking, bracing and brackets necessary for the installation of the lighting fixtures.

Provide ball type swivel hanger covers to permit 20 degree swing from vertical on all pendant fixture mountings. Swivel type hanger cover shall be equal to Appleton SHC 8458R.

17. All lighting fixtures, including glassware, lamps, reflectors and other accessories shall be thoroughly cleaned prior to final acceptance of the work.

18. Lighting fixtures shall be supported independently of ceiling structures not capable of supporting the fixture.

19. Supports for fluorescent fixtures shall have a minimum capacity of 150 pounds and all parts of the support shall be arranged to prevent vibrating free. Fluorescent fixtures installed in exposed grid system ceilings shall have independent support on at least two opposite corners or as directed.

20. Fixtures installed in the mechanical room shall be located and the height adjusted on the job to clear all obstructions such as ducts, piping, bracing and supports. The drawings are based on typical equipment for these areas and hence the contractor is held responsible for

locating fixtures so that proper illumination will be obtained. Where the location of fixtures shown on the drawings must be radically changed, approval from the architect shall be obtained before the fixture is placed.

21. Fluorescent fixtures mounted on combustible tile shall be spaced 1½ inches from the tile with suitable spacers, unless the fixtures are approved for direct mounting on combustible material.

22. Where lighting fixtures are recessed into acoustical tile ceilings, the fixture housing shall be installed in the ceiling in place of a tile. There shall be no cutting of tile around fixture housings.

23. Outlet boxes shall be accurately placed, independently and securely fastened to the structure, and set so that the plaster ring will be flush with the finished surface of the ceiling or wall.

Approved bar hangers and/or 2-inch by 4- or 6-inch solid blocking shall be used to support outlet boxes in all furred ceilings and stud walls. Hangers for lighting outlets shall have adjustable studs. Hangers for wall outlets shall have stove bolts for attaching the box to the hanger.

Where ductwork or piping interfere with the location as shown for boxes, the contractor shall relocate these boxes as directed by the architect without additional cost to the contract.

24. For local switch outlets, 4-inch pressed steel boxes with switch rings shall be provided for one and two ganged.

For convenience outlets, 4-inch pressed steel boxes with switch rings shall be provided.

For telephone outlets, $4^{11}/_{16}$-inch-square by $2^{1}/_{8}$-inch-deep pressed steel boxes with single gang cover shall be provided.

25. Pull boxes shall be code gauge steel and shall be installed wherever required in order to facilitate the pulling in of wires or cables in the conduit system. All boxes shall be provided with removable covers secured with machine screws. The interior and exterior of the box and cover shall be finished with two coats of primer coat and one coat of standard gray enamel.

26. Outlet boxes subject to damp or exterior locations shall be cast metal with threaded conduit hubs and gasketed covers.

27. Switches and receptacles shall be securely fastened to outlet box covers. Where the outlet box switch rings are back of the finished surface, the switch or receptacle shall be built out from the switch ring with washers or other approved means so that it is rigidly held in position in the box. The "floating" of any switch or receptacle will not be accepted.

Plates shall be secured to switches and receptacles with screws provided with the plate. These screws shall be of the same finish as the plate. Care shall be taken in setting plates that they do not buckle and they shall be square and plumb.

The mounting height for switches shall be 52 inches from finish floor to the center of the switch.

The mounting height for convenience outlets shall be 18 inches from finish floor to the center of the receptacle.

The mounting height for telephone outlets shall be 18 inches from finish floor to the center of the outlet.

The mounting height for plugmold wireways shall be 60 inches from finish floor to the center of the wireway.

Clock hanger outlets shall be mounted at 84 inches from finish floor to the center of the outlet.

28. Plugmold shall be installed securely to the structure with all necessary fittings. The plugmold shall be installed level and care shall be exercised to preserve the factory finish. Outlets shall be spaced equally at ends of run.

29. Main switchboard interior and exterior surfaces shall be thoroughly cleaned before applying energy to the system. The board shall be set level and plumb. Shim spacers shall be used where necessary. The board shall be placed on an equipment pad 4 inches thick and sized the same as the depth and width of the board. The board shall be anchored securely to the equipment pad and at the top anchored securely to the structure.

30. Panelboards shall be securely attached to the structure. The interior and exterior of each panel shall be thoroughly cleaned and free of unused conduit openings. The circuit directories shall be neatly typed to indicate each circuit breaker service. Where the panelboard schedule indicates spare, do not type in the word spare, leave the line blank for future use. Install cover plates level and plumb.

Music and Page System
Function and Description. Provide a FM music and paging system complete with cable splitter, FM receiver, amplifier, paging microphone,

speakers and volume control as shown on the drawings.

The system shall have two amplification channels and shall provide uninterruped FM music only in the office areas and both music and paging in the work areas. Cable shall be connected to FM music system and also FM outlets where shown.

1. Page microphone shall be Turner 253 microphone located at the receptionist's desk. Microphone cable shall be Belden No. 8723.

2. Receiver and amplifier shall be Raymer 840-35. Connect output of receiver to music only office speakers. Provide Raymer 800-35 amplifier and connect output of this amplifier to the work area speakers.

3. Speakers shall be Altec 409-8D with Audio Transformer Specialties M-1002 transformer, Soundolier T195-8 box, 81-8 T-bar bridge and T730-8B grille. Volume controls shall be Soundolier VC-5K. Speaker cable shall be Belden No. 8205.

4. All wire and cable shall be installed in a conduit system as shown on the drawings.

5. All of the music and paging system shall be serviced as required for one year. Calls for service shall be answered and completed within four (4) hours.

(Text continued on page 114)

Project Selection Checklist

Job   Advertising Company   **Bid Date**   To Be Announced

Location   Industrial Park   **Estimator**   Ed

Financial

1. Approximate Cost of Work   _____
2. Bonding Required   Yes
3. Progress Payments   Monthly
4. Retention   10 %
5. Delay Penalties $500/day
6. _____
7. _____

Project Type

1. Residential   _____
2. Commercial   X
3. Industrial   _____
4. Institutional   _____
5. Underground   _____
6. Overhead   _____
7. Waterfront   _____
8. High Voltage   _____
9. Communications   _____
10. _____
11. _____
12. _____

Bid Documents

1. Complete Plans   Elec.
2. Complete Specs.   Elec.
3. Reduced Plans   No
4. Plan Deposit   No
5. Public Bid   No
6. Sublisting   Yes
7. Prequalify   No

Basic Factors

1. Firm Price   X
2. Negotiated   _____
3. Special Equipment   _____
4. Construction Time   180 Cal days
5. Adequate Labor   Yes
6. Adequate Equipment   Yes
7. Adequate Tools   Yes
8. Site Conditions   Good
9. Unusual Problems   None
10. _____
11. _____
12. _____

Scope of Work

Estimate No. M-351

Job Advertising Company   **Bid Date** To Be Announced
Location Industrial Park   **Estimator** Ed
Bids To Architect   **Location** Hill Bldg.   **Time** 2:00 p.m.

1. Design Team
Architect	Hill & Hill
Engineer	Davis & Assoc.
Agency	N/A
Owner	Advertising Co.

2. Construction
Building	6,000 S.F.
Walls	Block/Gyp.
Ceilings	Exposed Grid
Floors	Carpet/Concrete

3. Quotations
Switchgear	X
Generator	N/A
Alarm Systems	N/A
U.F. Duct	N/A
Communication Systems	Music/Paging
Cable Tray	N/A
Fixtures	X
Telemetry	N/A

4. Specified Items
Conduit	GRS, EMT, PVC
Wire	THW, THHN, XHHW
Switches	1221-I
Receptacles	5242-I
Dimming Equipment	N/A
Motor Control	N/A
Manholes	N/A
Concrete	3,000 PSI

5. Related Work
Temporary	By G.C.
Control Wiring	By Mech.
Starters	X
Painting	By Others
Service Cable	By Utility Co.
Pole Bases	N/A

6. Site Conditions
Excavation	Trenching
Access	Clear
Utilities	Avail.
Security	By G.C.
Pave Cutting	N/A
Pave Patch	N/A
Demolition	

ELECTRICAL SYMBOL LIST

(A)	LIGHTING FIXTURE TYPE
S	SINGLE POLE SWITCH
ST	MANUAL MOTOR STARTER
⊖	DUPLEX RECEPTACLE
⊕	DOUBLE DUPLEX RECEPTACLE
⊖GFI	DUPLEX WITH GROUND FAULT INTERRUPTER
⊙	DUPLEX FLUSH FLOOR
◁	SPECIAL PURPOSE RECEPTACLE, SIZE AS SHOWN
◔	CLOCK HANGER RECEPTACLE
▭▬▭	PLUGMOLD
Ⓢ	SPEAKER, FLUSH
⑤	SPEAKER, SURFACE
◀	TELEPHONE OUTLET
◀	TELEPHONE, FLUSH FLOOR
☐	E.X.O. DISCONNECT SWITCH
F̄	E.X.O. FUSED
⊠	CONTROLLER, BY OTHERS, INSTALLED BY ELEC.
▣	PUSH BUTTON STATION
T̄	TRANSFORMER
☐○	4" BELL 12 VOLT
(EF)	EXHAUST FAN
S.P.	SUPPLY FAN
⊠	COMBINATION STARTER, BY ELEC.
▬▬	ELECTRICAL PANEL
▱	TERMINAL CABINET
W.P.	WEATHERPROOF
T.T.B.	TELEPHONE TERMINAL BACKBOARD
———	CONDUIT CONCEALED, ONE CIRCUIT
—⫴—	CONDUIT, HASH MARKS INDICATE NUMBER OF CIRCUITS
— — —	CONDUIT, UNDERGROUND
– – –	CONDUIT EXPOSED
—T—	TELEPHONE CONDUIT, MINIMUM SIZE 3/4"
—S—	SOUND CONDUIT, MINIMUM SIZE 1/2"

ADVERTISING CO.	
ELECTRICAL	E1
SYMBOL LIST	

LIGHTING FIXTURE SCHEDULE

	MANUFACTURER	CATALOG NUMBER	LAMP
Ⓐ	LITHONIA	2GP440 RN A12 ES	4-F40/SS
Ⓐ₁	LITHONIA	2GP240 RN A12 ES	2-F40/SS
Ⓑ	LITHONIA	4SG640 RN A12 ES	6-F40/SS
Ⓒ	LITHONIA	SC 240 ES	2-F40/SS
Ⓓ	PRESCOLITE	1015 F-3	100 A19
Ⓔ	PRESCOLITE	73222	2-F6T5
Ⓔ₁	PRESCOLITE	73322	2-F6T5
Ⓕ	MARCO	NKP22	100A21 I.F.
Ⓖ	LITHONIA	WA 296 A ES	2-F96T12/SS
Ⓗ	MARCO	B-113	2-F40/SS
Ⓙ	MARCO	M1-11	100A21 I.F.
Ⓚ	HOLOPHANE	2038-120-PHCA-UPH-35-120	150.H.P.S.
Ⓛ	APPLETON	G-50214	150A21 I.F.
Ⓜ	WESTINGHOUSE	FR-GSNG-W76A	250 H.P.S.
Ⓝ	LITHONIA	UN 240 ES	2-F40/SS

FIXTURE NOTES:

1. ALL FIXTURES AS SPECIFIED OR APPROVED EQUAL.
2. ALL FIXTURES TO BE COMPLETE AND WITH LAMPS.
3. ALL FIXTURE FINISHES TO BE MANUFACTURER'S STANDARD.

ADVERTISING CO.	
LIGHTING	E2
FIXT. SCHEDULE	

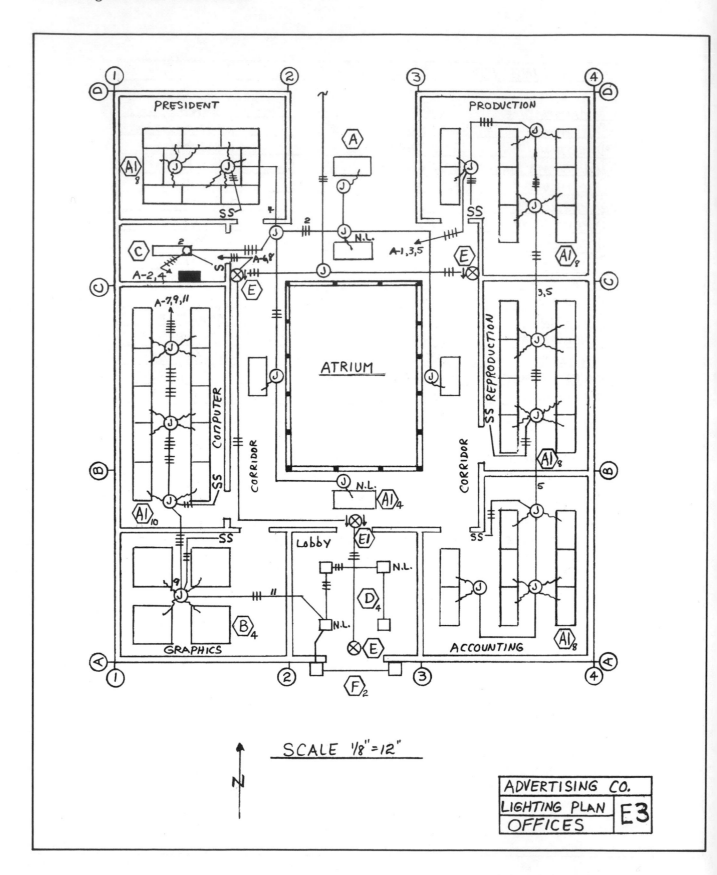

SCALE ⅛"=12"

N

ADVERTISING CO.
LIGHTING PLAN
OFFICES
E3

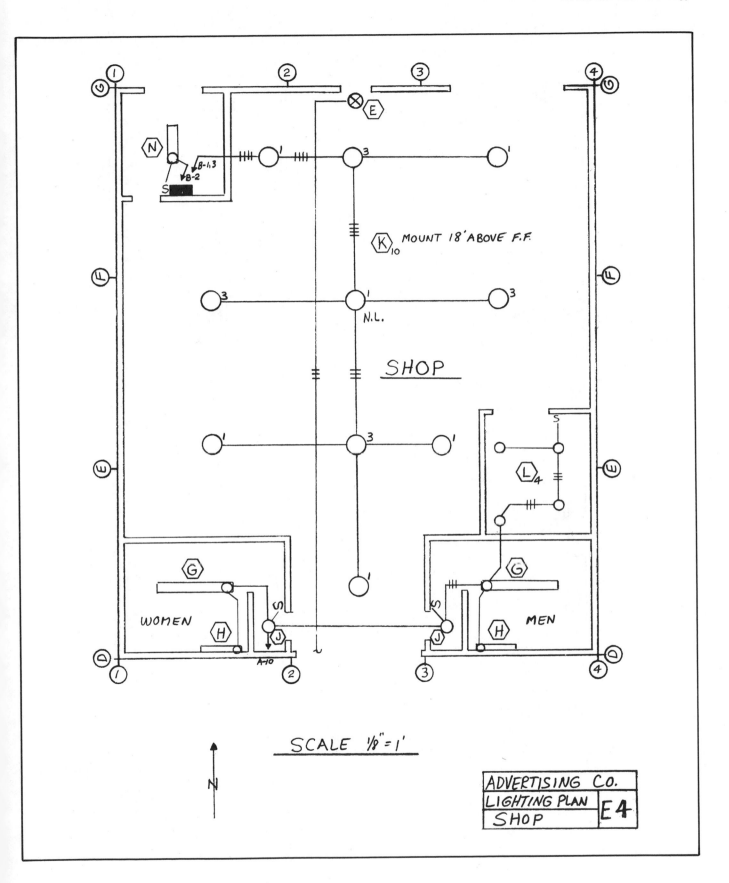

SHOP

MOUNT 18' ABOVE F.F.

N.L.

WOMEN

MEN

SCALE 1/8" = 1'

N

ADVERTISING CO.
LIGHTING PLAN
SHOP
E 4

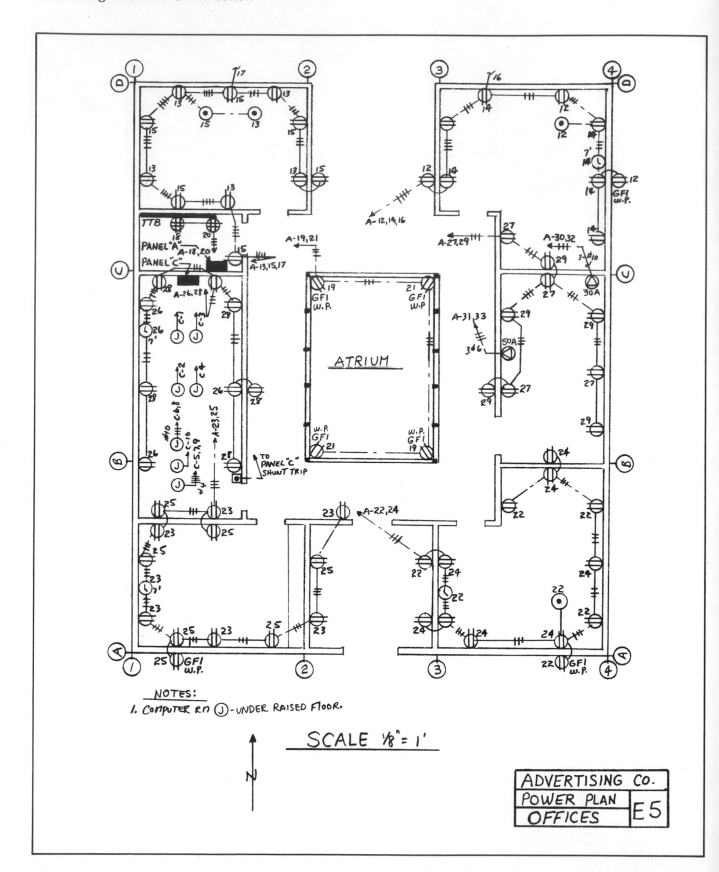

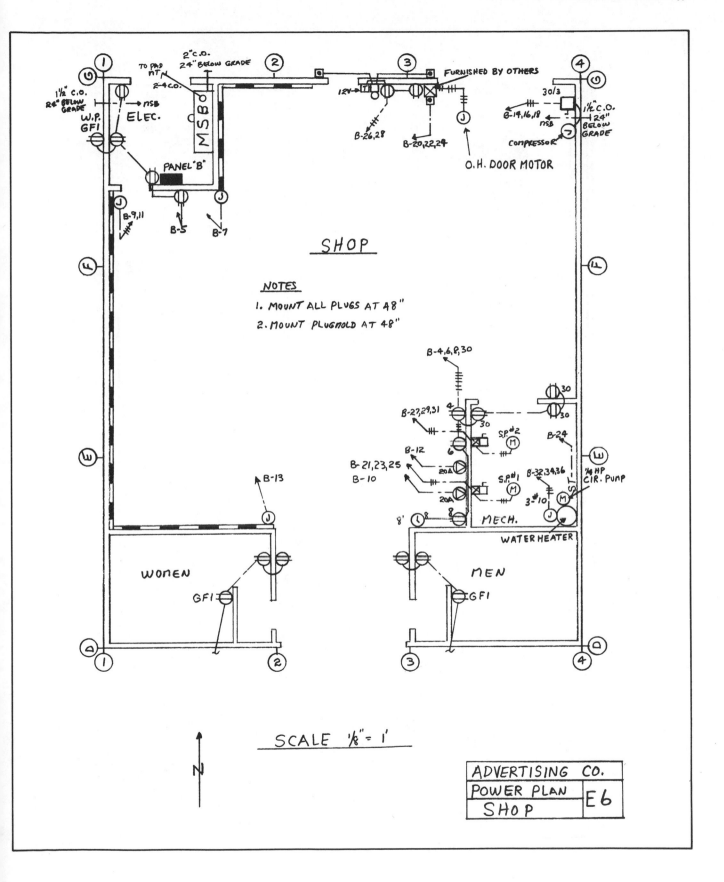

SHOP

NOTES
1. MOUNT ALL PLUGS AT 48"
2. MOUNT PLUGMOLD AT 48"

SCALE ⅛" = 1'

N

ADVERTISING CO.	
POWER PLAN	E6
SHOP	

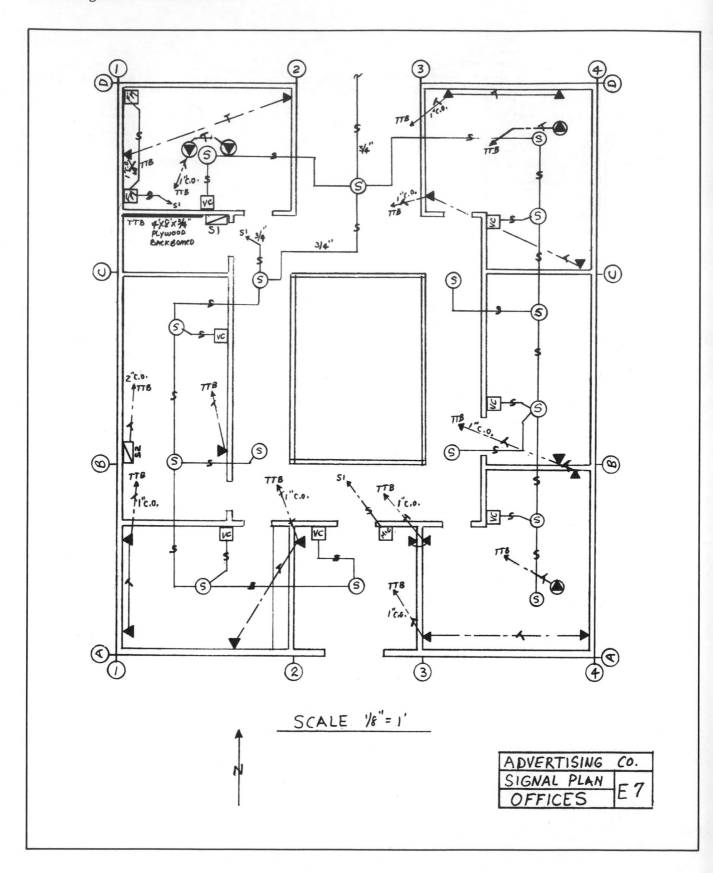

SCALE ⅛" = 1'

N

ADVERTISING CO.	
SIGNAL PLAN	E 7
OFFICES	

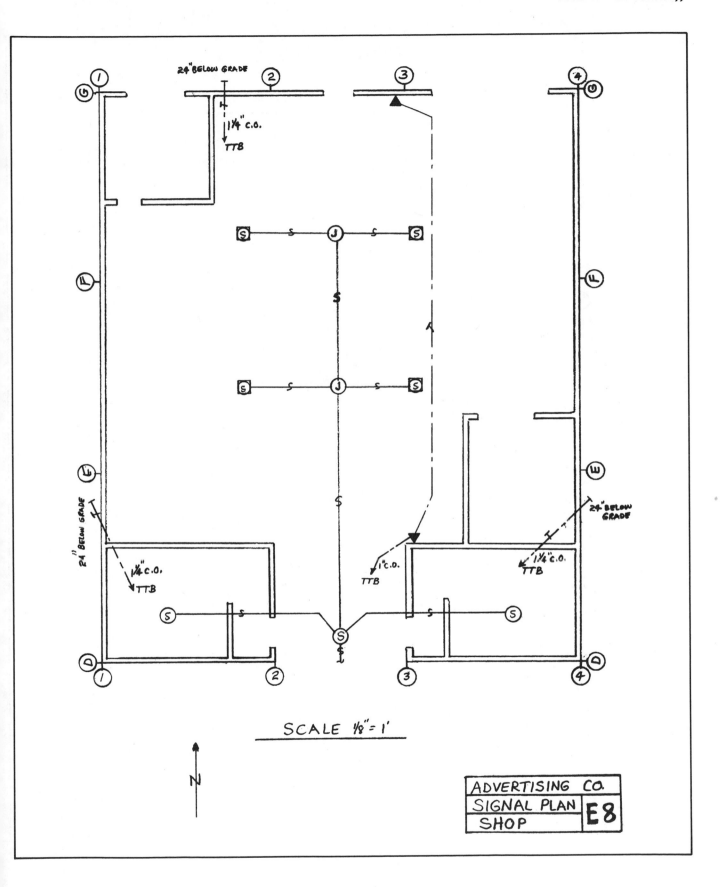

SCALE ⅛" = 1'

N

ADVERTISING CO.	
SIGNAL PLAN	E8
SHOP	

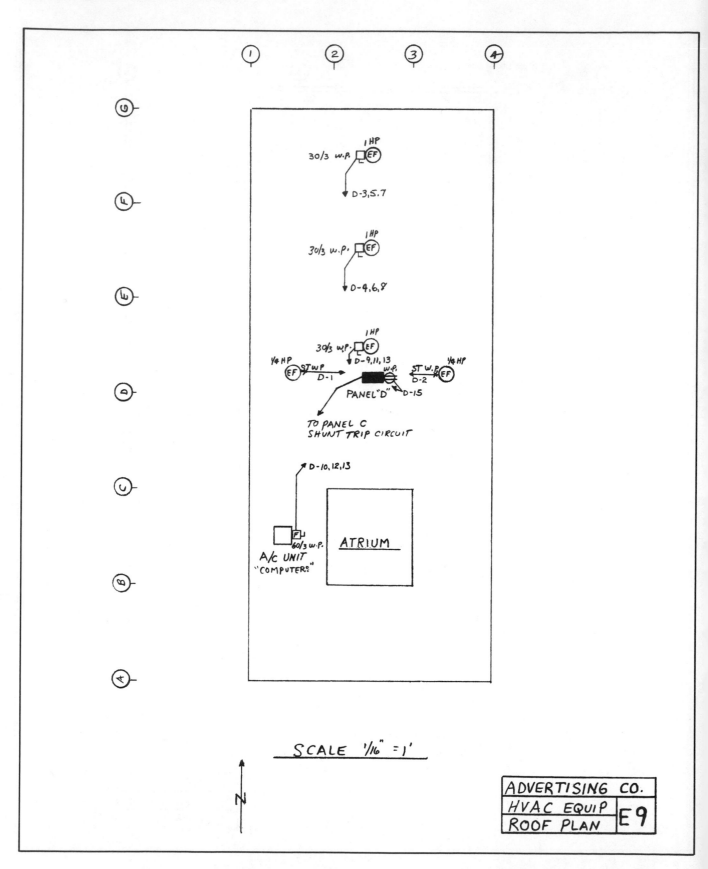

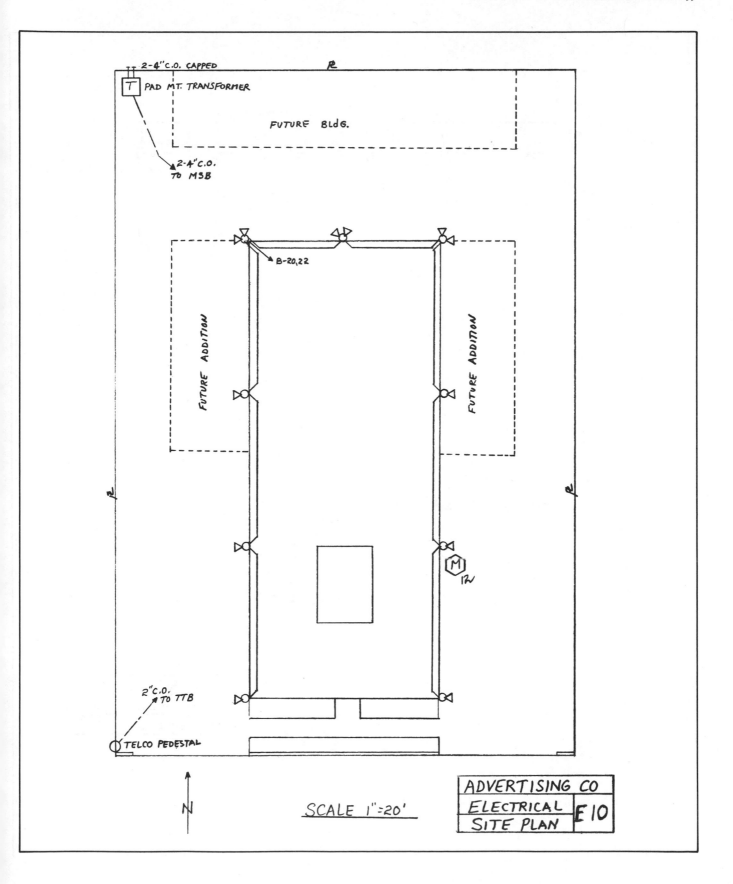

2-4"C.O. CAPPED

PAD MT. TRANSFORMER

FUTURE BLDG.

2-4"C.O.
TO MSB

B-20,22

FUTURE ADDITION

FUTURE ADDITION

M
12

2"C.O.
TO TTB

TELCO PEDESTAL

N

SCALE 1"=20'

ADVERTISING CO
ELECTRICAL
SITE PLAN   E10

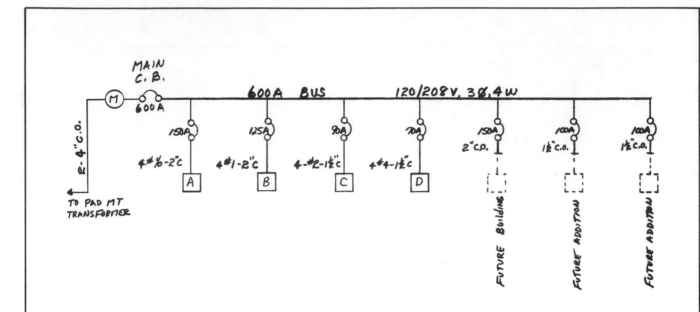

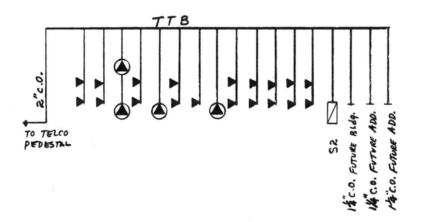

ADVERTISING CO.	
SINGLE LINE	E11
DIAGRAMS	

MOUNTING	SURFACE			PANEL	A			MAIN	M.L.O.	

120/208 VOLTS 3 PHASE 4 WIRE BUS 225A

#	Description	Type	A	B	C	C/B		C/B	A	B	C	Description	Type	#	
1	PRODUCTION	LTS	800			20		20	600			CORRIDOR	LTS.	2	*+
3	REPRODUCTION	LTS		800		20		20		800		PRESIDENT	LTS	4	
5	ACCOUNTING	LTS			800	20		20			50	EXIT	LTS	6	*
7	COMPUTER	LTS	1000			20		20	50			EXIT	LTS	8	*
9	GRAPHICS	LTS		1200		20		20		800		TOILET	LTS	10	
11	Lobby, ENTR.	LTS			600	20		20			900	PRODUCTION	PLUGS	12	+*
13	PRESIDENT	PLUGS	1080			20		20	1080			PRODUCTION	PLUGS	14	
15	PRESIDENT	PLUGS		1080		20		20		540		MEN	PLUGS	16	
17	WOMEN	PLUGS			540	20		20			360	TELCO	PLUGS	18	
19	ASTRIUM	PLUGS	360			20		20	360			TELCO	PLUGS	20	
21	ASTRIUM	PLUGS		360		20		20		1440		ACCOUNTING	PLUGS	22	
23	GRAPHICS	PLUGS			1260	20		20			1440	ACCOUNTING	PLUGS	24	
25	GRAPHICS	PLUGS	1260			20		20	900			COMPUTER	PLUGS	26	
27	REPRODUCTION	PLUGS		720		20		20		900		COMPUTER	PLUGS	28	
29	REPRODUCTION	PLUGS			900	20		30			2500	SPECIAL	PLUG	30	
31	SPECIAL	PLUG	4500			50		1	2500			SPECIAL	PLUG	32	
33	SPECIAL	PLUG		4500		1		20		1000		SPARE		34	
35	SPARE				1000	20		20			1000	SPARE		36	
37	SPARE		1000			20		20	1000			SPARE		38	
39	SPARE			1000		20		20		1000		SPARE		40	
41	SPARE				1000	20		20			1000	SPARE		42	
WATTS			10000	9660	6100				6490	6480	7250				

TOTAL WATTS	45980	=	45.98 KW =		130 AMPS	
	15326 X.25 ÷ 360		= LCL		11 AMPS	

PANEL NOTES:

1. * PROVIDE LOCKING DEVICE.
2. + CONTROLLED BY 2P TIME SWITCH.
3. PROVIDE NAMEPLATE ON PANEL TRIM, BLACK ON WHITE 3/8" LETTERS.

ADVERTISING CO	
PANEL	E12
SCHEDULE	

MOUNTING <u>SURFACE</u>   PANEL <u>B</u>   MAIN <u>M.L.O.</u>

<u>120/208</u> VOLTS <u>3</u> PHASE <u>4</u> WIRE BUS <u>225A</u>

Ckt	Description	A	B	C	Amp	Amp	A	B	C	Description	Ckt	
1	SHOP LTS	1050			20	20	100			ELEC RM LTS	2	
3	SHOP LTS		700		20	20		500		DISPENSER PLUG	4	
5	ELEC. RM PLUGS			900	20	20			500	DISPENSER PLUG	6	
7	PLUG MOLD	1200			20	20	500			DISPENSER PLUG	8	
9	PLUG MOLD		1200		20	20		500		DISPENSER PLUG	10	
11	PLUG MOLD			1200	20	20			500	DISPENSER PLUG	12	
13	PLUG MOLD	1200			20	20	300			COMPRESSOR 1/2 HP	14	
15	O.H. DOOR 1 HP		600		20			300			16	
17				600					300		18	
19			600			20	1500			YARD LTS	20	*+
21	SUPPLY FAN 2 HP		1200		20			1500			22	*+
23				1200		20			200	CIR. PUMP	24	
25			1200			20	200			SHIPPING PLUG	26	
27	SUPPLY FAN 2 HP		1200		20	20		200		SHIPPING PLUG	28	
29				1200		20			400	MECH. PLUGS	30	
31			1200			30	2500			WATER HEATER	32	
33	SPARE		1000		20			2500			34	
35	SPARE			1000	20				2500		36	
37	SPARE	1000			20	20	1000			SPARE	38	
39	SPARE		1000		20	20		1000		SPARE	40	
41	SPARE			1000	20	20			1000	SPARE	42	
WATTS		7450	6900	7100			6100	6500	5400			

TOTAL WATTS   39,450 = 39.45 KW =   115 AMPS

13,150 × .25 ÷ 360 = LCL   9 AMPS

<u>PANEL NOTES:</u>

1. * PROVIDE LOCKING DEVICE.
2. + CONTROLLED BY 2P TIME SWITCH.
3. PROVIDE NAMEPLATE ON PANEL TRIM, BLACK ON WHITE 3/8" LETTERS.

ADVERTISING CO.	
PANEL	E13
SCHEDULE	

MOUNTING **SURFACE** PANEL **C** MAIN **90A** SHUNT TRIP

120/208 VOLTS **3** PHASE **4** WIRE BUS **100A**

Ckt	Description	A	B	C	Brkr	Brkr	A	B	C	Description	Ckt
1	KEYPUNCH	1000			20	20	500			PRINTER	2
3	KEYPUNCH		1000		20	20		500		PRINTER	4
5	COMPUTER			6000	70	30			500	PROCESSOR	6
7		6000					500				8
9			6000			20		500		DOODLER	10
11	SPARE			1000	20	20			500	SPACE	12
	WATTS	7000	7000	7000			1000	1000	1000		
	TOTAL WATTS	24000	=	24 KW	=	70 AMPS					
		8000 X .25 ÷ 360	=	LCL	6 AMPS						

MOUNTING **SURFACE** PANEL **D W.P.** MAIN **70A** SHUNT TRIP

120/208 VOLTS **3** PHASE **4** WIRE BUS **100A**

Ckt	Description	A	B	C	Brkr	Brkr	A	B	C	Description	Ckt
1	TOILET E.F. 1/4HP	200			20	20	200			TOILET E.F. 1/4HP	2
3	SHOP E.F. 1HP		600		20	20		600		SHOP E.F. 1HP	4
5				600					600		6
7			600					600			8
9	SHOP E.F. 1HP		600		20	50		4500		COMPUTER A/C	10
11				600					4500		12
13			600				4500				14
15	ROOF PLUG		200		20	20		1000		SPARE	16
17	SPARE			1000	20	20			1000	SPARE	18
	WATTS	1400	1400	2200			5300	6100	6100		
	TOTAL WATTS	22500	= 22.5 KW	=	65 AMPS						
		7500 X .25 ÷ 360	=	LCL	5 AMPS						

PANEL NOTES:

1. PROVIDE SHUNT TRIP MAIN CIRCUIT BREAKERS FOR COMPUTER SHUT DOWN.

2. PROVIDE NAMEPLATES ON PANEL TRIM, BLACK ON WHITE 3/8" LETTERS.

ADVERTISING CO	
PANEL	**E14**
SCHEDULES	

Addendums

Addendums are written modifications or interpretations to the bidding documents before the bids are received. The modifications can be in addition to the original requirements or they can be deletions, clarifications or corrections. All addendums are written instructions and may include sketches or changed plans.

On occasion, the architect may issue more than one addendum to a set of bidding documents during the bid time. All of the addendums will become part of the contract documents after the contract has been executed.

When the architect or owner decides to extend the bid date for convenience, an addendum is issued simply stating the new bid date.

The commercial take-off exercise in this chapter has an addendum. The addendum lists changes that must be included in the bid price.

The addendum was received prior to the original take-off. Mark the changes on the drawing where the addendum directs. All of the changes can be covered as the take-off progresses.

If the addendum had been received after the original take-off was completed, then a separate take-off of just the changes could be made and the change cost either added or deducted from the original estimated cost.

Many times addendums are prepared from questions received from estimators when they find problems during the take-off. The problems must be called to the attention of the architect far enough in advance of the bid date to allow for the addendum. It is far better to have a firm understanding of the contract requirements before bid time and to also make sure that your competitors include any additional costs in their bid.

An example would be if the engineer having made a mistake and specifying a service switchboard that was too small for the building load. You discover the problem and know that corrections must be made. A change order later on after the contract is executed would delay the job and probably would not be worth the paperwork. The best thing to do would be to call the engineer and discuss the size of the switchboard. He may want to issue an addendum through the architect and be sure that all of the bidders correct the plans and price the correct size switchboard.

The following is addendum number 1 for the commercial take-off:

ADDENDUM NUMBER 1

ADDENDUM NUMBER 1

TO THE

SPECIFICATIONS AND DRAWINGS

FOR THE CONSTRUCTION OF

ADVERTISING COMPANY

Bid Opening:

The Date, Time and

Place shall **remain**

as specified on the

BID FORM

Make all revisions to the specifications and to the drawings as listed herein. All specification sections, drawings and details shall remain unchanged except sections, drawings and details that are added hereto, revised, deleted or clarified by this addendum. Attach this addendum to the specifications and acknowledge receipt of this addendum on the BID FORM. List the addendum number and list the date the addendum was received.

The addendum consists of three (3) pages.

ADDENDUM NUMBER 1   **ADVERTISING COMPANY**

16A   2.5   At the end of the paragraph add the following: "Provide a 6 inch diameter by 12 inch deep concrete marker directly over each underground capped conduit. Indicate on the marker face the exact depth of the conduit."

16A   2.6   At the end of the paragraph add the following: "Flexible Conduit shall not be used for general wiring."

16A   2.25   At the end of the paragraph add the following: "Label all pull boxes and junction boxes as to the circuit number."

16A   2.27   At the end of the section add the following: "The mounting height of convenience outlets in the Shop area, Mechanical and Electrical rooms shall be 52 inches above finish floor to the center of the device."

Drawing E4   **Shop Area** Overhead conduits shall be exposed maintaining an unobstructed clearance of 20 feet above finish floor.

Drawing E4   **Mechanical and Electrical Rooms** Conduits shall be exposed.

ADDENDUM NUMBER 1   **ADVERTISING COMPANY**

Drawing E5   **Computer Room** The Shunt Trip push button station shall be equal to Allen-Bradley 800T-D6LO mounted in a 800H-IHZ4R enclosure. Engrave the enclosure cover "SHUT DOWN" in ½ inch red filled letters. Mount at 60 inches above finish floor to the center of the push button.

Drawing E6   **Shop Area** Add 10 feet of Plugmold at column line 4, north from column line F. Connect to circuit B-33.

Drawing E8   **Shop Area** Overhead conduits shall be exposed maintaining an unobstructed clearance of 20 feet above finish floor.

The Take-off Procedure
Lighting Fixtures

Begin your take-off with the lighting fixture count. Write the number of the estimate in the upper right-hand corner of Work Sheet Number 1. List the various lighting fixtures by notation in a column at the left of the work sheet. These fixtures and their notations are found on the Lighting Fixture Schedule (Sheet E2). Watch for notations and special requirements that may affect the lighting fixtures. Next to the left column on your work sheet, list the first drawing that contains lighting fixtures. In our example, this is Sheet E3. Count each fixture in the drawing and record the totals on the work sheet. As you count the fixtures on the drawing, mark them with a colored pencil. This will keep you from counting a fixture twice.

When all fixtures in the drawing have been marked and counted, start a new column on the work sheet for the next drawing and repeat the counting procedure. Do this for each drawing that has lighting fixtures.

Next, list the lamp sizes on the work sheet. Lamp sizes and quantities are usually found on the Lighting Fixture Schedule. Multiply the number of lamps indicated for each fixture by the total number of fixtures. Then total the figures for each.

After the lighting fixtures and lamps have been taken off, number the work sheet at the top of the page as shown and place it in the estimate folder for safe keeping.

Your suppliers may want to know the number of fixtures so they can begin lining up the fixtures you need. It's to your advantage to give them approximate numbers. But don't pin yourself down to an exact count yet. Your lighting fixture count and in fact all material quantities are still approximate at this point. Give them only round numbers. They can get by with approximate counts at this stage.

Wiring Devices

Write the estimate number at the top of Work Sheet Number 2. On this work sheet you'll record the number of wiring devices, such as lighting switches, duplex receptacles and special outlets. List the actual counts in the left column of the work sheet. The symbols are found on the Electrical Symbol List (Sheet E1).

At the top of the work sheet list the first drawing containing wiring devices. In this case, it's Sheet E3. Check the drawing for special notations and requirements. Count each device, marking it with a colored pencil as before. Count and mark each device shown. Sometimes you may find devices on the drawing that are not on the Electrical Symbol List. When this happens, write the symbol on the work sheet and determine its use and size later.

When all devices have been counted and marked and the totals written on Work Sheet Number 2, start on the next drawing that has wiring devices. Write the drawing number on the work sheet and start counting and marking. Continue this procedure until all drawings have been checked and the totals written on the work sheet. If you run out of room on Work Sheet Number 2, continue on Work Sheet Number 3, as in the example.

Conduit, Connectors and Wire

On Work Sheet Number 4 record total conduit footage, total number of connectors and total footage of wire. Check the drawings for types and sizes of conduit. On the work sheet list the most common type of conduit first. (In our example, this is EMT conduit.) Now break down the conduit by size. Start with the smallest size first, as specified in the plans. The smallest conduit used on this job is 1/2 inch. For larger conduit check the Single Line, shown on Sheet E11. Larger conduits are usually used for panelboard feeders.

Measure conduit runs with a rotometer or a map measure. Note the plan scale. Here, the scale is 1/8 inch equals 12 inches.

Begin with the first plan drawing, in this example, Sheet E3. Start by measuring the 1/2-inch EMT conduit containing two No. 12 wires. Measure and mark all conduit lines in the drawing containing 2 wires. As each length of conduit is measured between fixtures, boxes or devices, count the 1/2-inch EMT connectors. Use a counting device to make an accurate count. When you've measured and marked all of the 1/2-inch EMT conduit with two No. 12 wires, record the total conduit footage, number of connectors and the total footage of wire on the work sheet. The wire should be at least double the conduit length. Allow extra wire for connections in fixtures, junction boxes, devices and panelboards. List the wire on a separate work sheet. In the sample it is on Work Sheet Number 6.

When measuring conduit from a ceiling-mounted fixture, junction box or device to a wall-mounted device, junction box or panelboard, allow for enough conduit to run down the wall to the conduit connector.

Next measure and mark all 1/2-inch EMT conduit containing three No. 12 wires. After listing the totals on the work sheets, measure and mark all of the conduit containing four No. 12 wires.

To find the maximum number of wires in a conduit, see Tables 1, 2, 3a, 3b and 3c in the National Electrical Code. Use Table 1 to determine the allowable percentage of conduit fill. Usually it's very difficult to install the maximum number of wires listed in Tables 3a, 3b, and 3c. Most electricians won't even try to pull the numbers listed in those tables.

It's wise to increase the conduit size after reaching about 80% of the allowable number of wires. You'll waste too much labor trying to pull more than that.

When all conduits have been measured, marked, and recorded on the work sheet, find the total of each type of conduit used. This is done by adding the totals for each drawing and placing these figures in a separate column on the work sheet.

Read all notes on the drawings, and mark each note after the related work has been recorded on the work sheets. Some notes may refer to other drawings or to requirements in other sections of the project. Check them thoroughly and allow for all installation material and labor. Also check each detail on the drawings.

Boxes and Covers

On Work Sheet Number 7 list the covers and plaster rings for the specified boxes. Each wiring device and most of the lighting fixtures require some type of box and cover.

Count and mark all covers and rings, and list the quantities for each drawing. The number of boxes can be determined later when transferring the take-off to the pricing sheets.

Watch for special boxes such as surface-mounted cast metal that may be specified for junction boxes or for wiring devices. List those boxes and indicate the size of the required conduit hub. A cast box may need more than one conduit hub. Check a catalog such as Crouse-Hinds or Appleton, and become familiar with the cast box configurations.

Sometimes the drawings indicate special

floor-mounted boxes. Check the specifications first to determine the type of box required. Check the drawing for notes that will explain the special box. If no information is available, make a reasonable guess on the use intended. Check the purpose of the special box, if possible. But be sure an allowance is made.

Be aware of changes in wiring methods. And take note of new restrictions. There are many different classes of wiring systems. Hazardous locations are noted by class and division in the National Electrical Code, Article 500.

Total the Work Sheets

When all drawings have been marked and all material and work have been listed on the work sheets, check the addenda once again to be sure that all changes have been included.

Start a column on each work sheet for the totals. Adding across, determine the total number of materials used for each type of material. Place these totals in the "total" column on the right.

When all totals have been listed, the take-off is finished and can be transferred to the pricing sheets.

After you've completed the transfer, staple the work sheets together and place them in the estimate folder. You may need them later to trace certain materials. Be sure to check off each item as it is transferred.

Transfer to the Pricing Sheets

Start the transfer from the work sheets by listing the lighting fixture types on the pricing sheets. Then list the types of lamps required. It's good practice to list the lamps immediately after the fixtures. The supplier may include all or just part of the lamps in the pricing of the fixtures. Also, listing the fixtures and lamps together permits easier cross checking for verification.

Now, list the conduit and fittings, beginning with the smallest size of the most common type of conduit. List the connectors or terminations for EMT conduit, allowing one coupling for each ten feet of conduit. PVC and Rigid have a coupling on one end of the ten-foot length. Allow two additional couplings for each factory 90-degree elbow. Usually smaller elbows are made at the job site. Those larger than one inch are purchased. On projects requiring many large elbows, a special hydraulic conduit bender may be useful. It can make a bend in a

Work Sheet

#1   Estimate No.: M 351

TYPE of FIXTURE	E3	E4	E10	TOTAL		Lighting Fixtures / Lamps 100 A19 CLEAR	100 F21	150 A21	150 MHS	250 HPS	F67S	F40 WW/SS	F96T2WW
A	1			1								4	
A1	46			46								92	
B	4			4								24	
C	1			1								2	
D	4			4		4							
E	3	1		4							8		
E1	1			1							2		
F	2			2			2						
G		2		2									4
H		2		2								4	
J		2		2			2						
K		10		10					10				
L		4		4				4					
M			12	12						12			
N		1		1								2	
					TOTAL	4	4	4	10	12	10	128	4

Work Sheet

#2

Estimate No.: *M351*

WIRING DEVICES

	E3	E4	E5	E6	E7	E8	TOTAL
S	1	4					5 ✓
SS	6						6 ✓
⊖			61	16			77 ✓
⊖ GFI W.P.			7	1			8 ✓
⊕			2				2 ✓
⊙			4				4 ✓
Ⓛ			4	1			5 ✓
△ 30			1				1 ✓
△ 50			1				1 ✓
SHUNT TRIP PB			1				1 ✓
⊖ GFI				2			2 ✓
20				2			2 ✓
PLUGMOLD 1 CIR.				36'			36' ✓
" 2 CIR.				48			48' ✓
" ADAPTER				4			4 ✓
" END				8			8 ✓
" ELBOW				1			1 ✓
12V				1			1 ✓
12V				1			1 ✓
12V				2			2 ✓
ST				1			1 ✓
30/3				1			1 ✓
SIZED 30/3				2			2 ✓
O.H. DOOR CONTROL				1			1 ✓
P.B.				1			1 ✓
COMPRESSOR				1			1 ✓
CIR. PUMP				1			1 ✓
SUPPLY FANS				2			2 ✓
WTR. HTR				1			1 ✓
O.H. DOOR				1			1 ✓
◄					17	2	19 ✓
◓					4		4 ✓
Ⓢ					16	3	19 ✓
DJ					1		1 ✓

Work Sheet

#3 Estimate No.: M351

	E1	E8	E9	E12	E13		TOTAL
Fm	2						2✓
mc	1						1✓
◻	2						2✓
4'X8'X3/4" TTB	1						1✓
Ⓢ		4					4✓
⊖ WP			1				1✓
STWP			2				2✓
⊡ 30/3 WP			3				3✓
F 60/3 WP			1				1✓
40A FUSES			3				3✓
E.F. ¼ HP			2				2✓
" 1 HP			3				3✓
A/C UNIT			1				1✓
T.S. 2 POLE				1	1		2✓

Work Sheet

#4   Estimate No.: M351

CONDUIT / FITTINGS	E 3	E 4	E 5	E 6	E 7	E 8	E 9	E 10	E 11	TOTAL
1/2" EMT	550	420	200	90	290	130	190	320		2190
3/4"	20		30	20	80					150
1"			3		5		50			58
1¼"			30							30
1½"									90	90
2"					4				16	20
1/2" CONN	76	52	124	47	48	16	14	18		395
3/4"	4		26	4	19	2				55
1"			4		5	2	2			13
1¼"			2							2
1½"									4	4
2"					2				4	6
1¼" Elb.			2							2
1½"									3	3
2"									3	3
1/2" PVC			310	380	50					740
3/4"			120	100	220	50				490
1"			40		320	40				400
1¼"						180				180
1½"				60					75	135
2"				10	25			70	75	180
4"								150		150
1/2" FA			45	30	4					79
3/4"			4	4	12	2				22
1"			2		17	2				21
1¼"										-0-
1½"				4					2	6
2"				2	2				2	6
1/2" TA			5							5
3/4"					4					4
1"					1					1
1/2" Elb.			50	30	4					84
3/4"			2	2	16	2				22

Work Sheet

#5

Estimate No.: **M351**

	E5	E6	E7	E8	E9	E10	E11	TOTAL	
CONDUIT / FITTINGS									
1" Elb.	2		18	2				22	
1¼"				3				3	
1½" PVC Elb.		2						2	
2"		1	2			2	2	7	
4"						6		6	
4" 45°						2		2	
1¼" CAP				3				3	
1½"		2						2	
2"		1						1	
1¼" COUP								-0-	
1½"		2						2	
2"		1						1	
4"						8		8	
½" RIGID	140	40	10					190	
3/4"			5					5	
1"			2					2	
½" TERM	56	23	3	1				83	
3/4"			5					5	
1"			2					2	
2 BASE SPACERS						14		14	
4						30		30	
TRENCH 10X24		15	25	15		70		125	
20X40						75		75	
MARKERS		3		3				6	
½" ROOF FLSH.					5			5	
3/4"					1			1	
PITCH POCK 6X10X4					1			1	
4" PVC CAP						2		2	
Pd. mt. SLAB						1		1	
CONCRETE						6		6	

Work Sheet

#6

Estimate No.: M351

WIRE

	#3	#4	#5	#6	#7	#8	#9	#10	#11	TOTAL
#12	1760	1040	2640	2120			490	640		8690
10			300	200						500
8						160				160
6			170							170
4			120						360	480
2									360	360
1									80	80
1/0									360	360
STRING					600	280				880

Work Sheet

#1   **Estimate No.:** M 351

	E3	E4	E5	E6	E7	E8	E9	E10	TOTAL	BOXES/COVERS
4/S.P. RING 5/8	11	9							20	
3/0 EXT. "	2								2	
4/S SW 5/8 "	1	4	36	19	13				73	
" " 2GANG 5/8	6		2						8	
" BLANK	18		6	2		2		9	37	
4"/16 "	3			1					4	
4/S SW RING 1"x4			32	11	6	1			50	
4"/16 SW RING 5/8			1		4	1			6	
CAST BOX FS			4				3		7	
4"/16 SW RING 1"x4					3				3	

Work Sheet

#8

Estimate No.: $M351$

	E 1	E 8	E 9	E 12	E 13		TOTAL	
FM	2						2 ✓	
MC	1						1 ✓	
◢	2						2 ✓	
4'X8'X3/4" TTB	1						1 ✓	
Ⓢ		4					4 ✓	
⊖ WP			1				1 ✓	
STWP			2				2 ✓	
☐ 30/3 WP			3				3 ✓	
F 60/3 WP			1				1 ✓	
40A FUSES			3				3 ✓	
E.F. 1/4 HP			2				2 ✓	
" 1 HP			3				3 ✓	
A/C UNIT			1				1 ✓	
T.S. 2 PLE				1	1		2 ✓	

ten-foot length of conduit, thus saving the cost of the elbow, the couplings and the labor to install them. Be sure to include labor for field bending the larger elbows.

List all boxes, wire, devices, safety switches, motor control devices, hookups, special material, trenching and concrete for the job.

After transferring all material and work items to the pricing sheets, list the switchgear. Then staple the pricing sheets together and number them.

Pricing and Labor

Try to anticipate the quoted prices for switchgear, fixtures and other items. This will help you become more familiar with the price levels. You may even be able to detect supplier errors. Eventually, your guesses will fall pretty close to the quoted prices.

Some contractors use purchasing agents for pricing parts of the estimate. Don't pass your responsibilities to someone else. The purchasing agent doesn't know anything about the construction schedule, site location or the many other factors that can affect material and equipment costs. You know the project best. If you use a purchasing agent, use him for the common bread and butter items only.

Basic pricing standards are used throughout the industry. First, there are unit prices or "each." When an item is priced or labored by "each," use the letter "E." On items extended by the "hundred," use the letter "C." Indicate thousand with the letter "M." For items covered in one lump sum, use "lot." Pavement patching is measured by the Square Foot (SF). Concrete and excavation are measured by the Cubic Yard (CY).

When all take-off items have been priced and labored, extend each item and total the pricing sheet. It's faster to list the unit costs and labor units throughout the estimate first, then go back and extend the whole package. It's also more accurate.

Check the Pricing Sheets

After the pricing sheets have been priced, extended and totaled, check them for errors. No matter how thorough you've been, there's still the possibility of error.

Don't check the estimate yourself. You may make the same mistake twice. Have someone else do it. Tell the person checking the estimate to check the figures that are right with a check

mark and mark any mistakes with a colored pencil. And be sure to tell the checker not to make any corrections. You're the only one who should make corrections.

When the checking is finished, go through the estimate and check the errors marked. Find out why the error was made. Then make the necessary corrections.

The Bid Summary

When you're sure the pricing sheets are correct, prepare the bid summary sheet. This becomes the estimate package index as well as the final estimate pricing page.

List each of the materials from the pricing sheet in sequence. Write down the total labor and material cost from the pricing sheet for each item. If you run out of room, combine like materials, such as conduit and fittings.

Total the indexed labor and material. To double check your bid summary sheet at this point, total the material at the bottom of each pricing sheet; this total should agree with the index total.

When the bid summary sheet is completed, staple it to the estimate package.

Telephone Quotations

Use a form to record all necessary data received from telephone quotations. A well-designed form will help you remember important information and helps you organize the quotes you get.

List the job name, the estimate number, and the date and time the quotation was received. Write the name of the company and of the person offering the quotation. Fill in the form as completely as possible, and make as many notes as you need.

Find out if the price includes tax, freight, and delivery to the job site. Ask about anticipated price increases or exclusions. If you see any problems, make note of them. All of these affect your cost.

After bidding a few jobs, you'll become familiar with the people who handle telephone quotations for the suppliers. Many suppliers will assign one person to your company to handle all of your bidding. Most furnish written quotations prior to bid time. These quotations usually list material or equipment, terms of sale and price. That's a big help because it serves as a check against the estimate. The written quotation may also list the quantities of lighting fixtures. This, too, helps you spot any discrepancies and make corrections.

On a larger job the supplier may send a manufacturer's representative to your office to do an equipment take-off or to explain material installation. This may be a big help in estimating labor for an item. Sometimes the rep can point out supporting materials for equipment that would otherwise be overlooked. The rep is also a good source of new product information and installation procedures.

Always be fair in dealing with those quoting. Don't give out their quotations to others. If you tell one supplier about the prices you've received from another supplier, you'll lose the trust the suppliers have in you. A reputation for fair dealing is one of your most important assets.

File each written or telephone quotation in the estimate folder. You'll need them when you write purchase orders for the material or equipment quoted.

The Spread Sheet

Not all suppliers will provide written quotations before the bid date. To prepare for the last minute dash at bid time, prepare a spread sheet to compare material and equipment quotes.

Write the name of the project and the estimate number on the spread sheet. List the types and quantites of lighting fixtures in the left-hand columns. As the quotations are received, write the name of the supplier and the salesperson at the top of the sheet.

Use the spread sheet to compare supplier prices. Having all quotes on one page makes it easier to spot errors. Inform a supplier if his price seems out of line. Don't mention other quotes; just give them the chance to correct the mistake.

A spread sheet also helps you keep track of who is quoting the job and who you might have missed. Bid time is usually very hectic and a supplier's price can easily be missed. It won't help to learn later that there was a better price that would have gotten you the job.

Completing the Estimate

The end sheet is usually completed on the day of the bid. Make sure the pricing sheets are correct and in order. You should have at least three quotations for each major item of material or equipment. Try to get as many

quotations as possible. Some manufacturers' representatives offer quotations directly to contractors, with the supplier's markup included. Some manufacturers sell directly to the contractor, without the supplier's markup.

In the index section of the end sheet there's a column for deducts. Sometimes you will have changes to the guesstimated price. Indicate additions with a "+" and deductions with a "—". List the adjusted price or difference between the spread sheet price and the indexed price in the deduct column. Total the column and enter that amount in the first subtotal box.

If a sales tax applies, calculate the amount for both the material and deduct columns. Add the tax to the first subtotal price and list the new total in the next subtotal box for each column.

Move over to the labor column. Many projects require additional manhours for supervision. Calculate the most likely amount of supervision that will be required and enter it as a percentage. List additional manhours calculated from that percentage.

Add the subtotal manhours and the additional supervision manhours. List the sum in the total hours box. Multiply the total hours by the hourly rate paid to the workers. List the base labor cost in the next subtotal box.

Fringe benefits include health and welfare, pensions and vacations. They can be calculated by a set percentage or by the actual hourly cost.

Labor taxes include Worker's Compensation, insurance, unemployment and social security. Again, the amount can be calculated by a percentage or by an hourly amount.

After calculating fringe benefits and taxes, add these amounts to the subtotal and list the figure in the next subtotal box.

Sometimes labor should be factored. Certain jobs require other than the average amount of labor units on the pricing sheets. Perhaps the job is in an area with adverse conditions or distractions. If so, factor the labor cost to include these conditions.

Write the total labor cost in the box on the left side of the end sheet.

Tools are a direct job cost. The ones to consider are the "expendable" tools such as ladders, drill motors and other small tools. The cost can be calculated as a percentage of labor. The sample end sheet for estimate M351 has a tool budget of about 6½ percent of the cost less fringe benefits and labor taxes. The amount is rounded off to the nearest hundred.

"Miscellaneous" is for material that will be needed in addition to that listed on the pricing sheets. The amount listed on estimate M351 is about one percent of the first material subtotal.

Permits and fees are entered in the next box. These costs vary with every authority around the country. Find out how permits and fees are calculated in your area and list that amount on the estimate.

Add up the figures for material and tax, labor, tools, miscellaneous, and permits and fees. List the subtotal in the next box.

"Subcontracts" is for quotations for services such as the music and paging system that is furnished and installed by the sound contractor. List the best price from the music and paging spread sheet in the subcontract box.

Travel expense varies from job to job. Most local jobs are not assessed travel expense. Estimate M351 is in town. List "N/A" in the box to indicate "Not Any."

Add the subcontracts to the subtotal and list the sum in the next subtotal box. Now bring down the subtotal from the deduct column and list the amount in the deduct box. Subtract, and list the net cost.

Charge overhead on every job, no matter how small your operations are. Preparing the estimate takes time and money. Payroll, purchasing, bookkeeping, office rental, and office equipment are all basic overhead costs. Establish a certain percentage to cover overhead. Calculate the amount and list the sum in the overhead box. Add that to the net cost. Then list the sum in the next subtotal box

Profit is the return on the money invested by your company. You have to establish your own profit goal. Naturally, there is more to profit than just including the amount you want to earn. Some electrical contractors charge more than 20 percent and still get enough jobs to stay busy.

No matter how large or small your profit

percentage is, be sure that it is really a true profit. "Profit" should not be an allowance for unexpected items. It isn't management's fee or money to cover any other cost. Your estimate covers all the costs. Profit should be strictly a return on the money invested by your company — like interest from a bank account. It shouldn't be used to cover any labor, material or other cost.

Use the contingency allowance to cover unforeseen costs such as material price increases. Suppliers usually quote prices based on delivery within a certain number of days. If that time passes, the price may be higher. Use the contingency column to cover any cost which could not be anticipated but which experience tells you may be incurred.

Many contractors and owners require a performance bond. Even if electrical costs are only about 10 percent of the total project cost, the contractor may require a performance bond. Include the bond cost here.

Finally, add the profit, contingency and bond to the last subtotal and list the sum in the selling price box. Many contractors round off the total.

Have someone review the bid summary calculations to be certain that there are no errors. If you've worked carefully and accurately, your estimate will be as nearly perfect as humanly possible.

The Bid Summary on page 131 is an index of all the pricing sheets for estimate no. M-351, pages 132 to 139. The takeoff has been checked by Sue and can now be put aside until the suppliers quote on the switchgear and fixtures.

On or about the day of the bid, the suppliers will call in the prices as indicated on the telephoned quotations forms found on pages 140 to 152. Spread sheets, pages 153 to 155 are used for comparing the prices that were quoted.

Finally, the bid summary, page 156, can be completed. The selling price might be adjusted slightly in preparation for calling out the bid to the General Contractors.

Bid Summary

Job: **ADVERTISING COMPANY**   Estimate No.: **M-351**
Location: **INDUSTRIAL PARK**   Bid Date: _____
Division: **16 A ELECTRICAL**   Time: **2:00 P.M.**
Estimator: **ED**   Checker: **SUE**

	Description	Material	Deducts		Labor
1.	SWITCHGEAR	5 000.00			52.65
2.	FIXTURES	8 278.00			78.00
	LAMPS	1 273.00			9.90
3.	CONDUIT/FITTINGS	1 225.00			211.66
4.	" "	331.00			26.45
5.	" BOXES	491.00			82.26
6.	WIRE	1 039.00			95.90
7.	DEVICES	2 287.00			65.50
8.	HOOKUP/TRENCH	1 174.00			43.35
	Sub Total	21 099.00			665.67

	Material			
Sales Tax ____%		Supervision %	
Sub Total		Total Hours	
Labor		Rate	
Tools		Sub Total	
Miscellaneous		Fringes	
Permits/Fees		Taxes	
Sub Total		Sub Total	
Subcontracts		Factor	
Travel Expense		Total	
Sub Total			
Deducts		Addendums	
Net Cost		Alternates	
Overhead		Job Duration	
Sub Total		Penalty	
Profit		Type Const.	
Contingency			
Bond			
Selling Price			

Pricing Sheet

Job: _____   Estimate No.: *M-351*

Work: _____   Sheet: _1_ of _8_

Estimator: *ED* _____ Checker *Sue* _____   Date: _____

Description	Qty.	Price	Per	Extension	Hours	Per	Extension	
MSB 120/208 .3∅ 4W	1			5,000.00		E	18.50	✓
METERING 600A BUS								
MAIN C/B 600/3								
1-70/3 C/B								
1-90/3								
2-100/3								
1-125/3								
2-150/3								
PANEL A SURFACE	1					E	14.00	✓
M.L.O. 225A BUS								
38-20/1 C/B								
1-30/2								
1-50/2								
PANEL B SURFACE	1					E	12.00	✓
M.L.O. 225A BUS								
25-20/1 C/B								
1-20/2								
4-20/3								
1-30/3								
PANEL C SURFACE	1					E	4.10	✓
90-A SHUNT TRIP 100A BUS								
1-20/1 C/B								
1-30/2 C/B								
1-70/3 C/B								
PANEL D W.P.	1					E	4.05	✓
70A SHUNT TRIP 100A BUS								
6-20/1 C/B								
3-20/3 C/B								
1-50/3 C/B								
Total				5,000.00	✓		52.65	✓

Pricing Sheet

Job: _____ Estimate No.: *M351*

Work: _____ Sheet: *2* of *8*

Estimator: _____*ED*_____ Checker _____*SUE*_____ Date: _____

Description	Qty.	Price	Per	Extension	Hours	Per	Extension	
FIXTURE A	1	—	E	45.00	—	E	1.00	
A1	46	34.00	E	1564.00 ✓	.75	E	34.50	✓
B	4	210.00	E	840.00 ✓	2.00	E	8.00	✓
C	1	—	E	48.30	—	E	.65	
D	4	45.00	E	180.00 ✓	.40	E	1.60	✓
E	4	45.10	E	180.40 ✓	.40	E	1.60	✓
E1	1	—	E	51.35	—	E	.40	
F	2	65.00	E	130.00 ✓	.40	E	.80	✓
G	2	82.70	E	165.40	1.25	E	2.50	✓
H	2	44.00	E	88.00 ✓	.65	E	1.30	✓
J	2	28.00	E	56.00 ✓	.50	E	1.00	✓
K	10	135.00	E	1350.00 ✓	1.00	E	10.00	✓
L	4	21.26	E	85.04 ✓	.50	E	2.00	✓
M	12	289.00	E	3468.00 ✓	1.00	E	12.00	✓
N1	1	—	E	27.00	—	E	.65	
				—			—	
100A19CL LAMPS	4	53.00	C	2.12 ✓	.05	E	.20	✓
100A 21	4	61.00	C	2.44 ✓	.05	E	.20	✓
150A21	4	77.00	C	3.08 ✓	.05	E	.20	✓
150 HPS	10	4001.00	C	400.10 ✓	.10	E	1.00	✓
250 HPS	12	4142.00	C	497.04 ✓	.10	E	1.20	✓
F6T5 WW	10	462.00	C	46.20 ✓	.05	E	.50	✓
F40 WW/SS	128	236.00	C	302.08 ✓	.05	E	6.40	✓
F96 T12 WW/SS	4	496.00	C	19.84 ✓	.05	E	.20	✓
Total				9551.39 ✓			87.90	✓

Pricing Sheet

Job: _____ Estimate No.: *M351*
Work: _____ Sheet: *3* of *8*
Estimator: ____*ED*____ Checker ___*SUE*___ Date: _____

Description	Qty.	Price	Per	Extension	Hours	Per	Extension	
1/2" EMT CONDUIT	2190	11.25	C	246.38	3.25	C	71.18	✓
3/4"	150	16.10	C	24.15	3.75	C	5.63	✓
1"	60	24.32	C	14.59	4.25	C	2.55	✓
1 1/4"	30	34.10	C	10.23	4.75	C	1.43	✓
1 1/2"	90	41.81	C	37.63	5.75	C	5.18	✓
2"	20	50.33	C	10.07	7.50	C	1.50	✓
1/2" CONNECTOR	395	22.00	C	86.90	.05	E	19.75	✓
3/4"	55	32.85	C	18.07	.06	E	3.30	✓
1"	13	54.90	C	7.14	.08	E	1.04	✓
1 1/4"	2	99.50	C	1.99	.10	E	.20	✓
1 1/2"	4	150.00	C	6.00	.10	E	.40	✓
2"	6	209.00	C	12.54	.15	E	.90	✓
1/2" COUPLING	200	28.65	C	57.30	.05	E	10.00	✓
3/4"	15	39.60	C	5.94	.06	E	.90	✓
1"	6	65.70	C	3.94	.08	E	.48	✓
1 1/4"	5	110.05	C	5.50	.10	E	.50	✓
1 1/2"	15	159.35	C	23.90	.10	E	1.50	✓
2"	8	216.40	C	17.31	.15	E	1.20	✓
1 1/4" ELBOW	2	170.00	C	3.40	.10	E	.20	✓
1 1/2"	3	214.14	C	6.42	.10	E	.30	✓
2"	3	350.29	C	10.51	.15	E	.45	✓
EMT HANGERS/STRAPS	LOT	—	—	40.00	—	—	8.00	✓
1/2" PVC CONDUIT	740	10.16	C	75.18	3.10	C	22.94	✓
3/4"	490	13.50	C	66.15	3.20	C	15.68	✓
1"	400	19.98	C	19.92	3.30	C	13.20	✓
1 1/4"	180	27.03	C	48.65	3.40	C	6.12	✓
1 1/2"	140	32.20	C	45.08	3.45	C	4.83	✓
2"	180	42.94	C	77.29	3.50	C	6.30	✓
2"	150	122.05	C	183.08	4.00	C	6.00	✓
Total				1225.26 ✓			211.66 ✓	

Pricing Sheet

Job: _____ Estimate No.: *M351*
Work:_____ Sheet: *4* of *8*
Estimator: *ED*   Checker *SUE*   Date: _____

Description	Qty.	Price	Per	Extension	Hours	Per	Extension	
½" PVC ELBOW 90°	84	.46	E	38.64 ✓	.05	E	4.20	✓
¾"	22	.51	E	11.22	.05	E	1.10	✓
1"	22	.80	E	17.60 ✓	.05	E	1.10	✓
1¼"	3	1.10	E	3.30	.10	E	.30	✓
1½"	2	1.52	E	3.04	.10	E	.20	✓
2"	7	2.22	E	15.54 ✓	.15	E	1.05	✓
4"	6	12.19	E	73.14 ✓	.20	E	1.20	✓
4" 45°	2	11.29	E	22.58 ✓	.20	E	.40	✓
½" FA	79	.18	E	14.22 ✓	.05	E	3.95	✓
¾"	22	.30	E	6.60 ✓	.05	E	1.10	✓
1"	21	.41	E	8.61 ✓	.05	E	1.05	✓
1½"	6	.56	E	3.36 ✓	.10	E	.60	✓
2"	6	.80	E	4.80 ✓	.15	E	.90	✓
½" TA	5	.18	E	.90 ✓	.05	E	.25	✓
¾"	4	.33	E	1.32 ✓	.05	E	.20	✓
1"	1	.41	E	.41 ✓	.05	E	.05	✓
1¼" CAP	3	.62	E	1.86 ✓	.10	E	.30	✓
1½"	2	.71	E	1.42 ✓	.10	E	.20	✓
2"	1	.35	E	.35 ✓	.10	E	.10	✓
4"	2	.74	E	1.48 ✓	.15	E	.30	✓
½" COUPLINGS	12	.15	E	1.80 ✓	.05	E	.60	✓
¾"	12	.18	E	2.16 ✓	.05	E	.60	✓
1"	12	.28	E	3.36 ✓	.05	E	.60	✓
1½"	2	.49	E	.98 ✓	.10	E	.20	✓
2"	1	.70	E	.70 ✓	.10	E	.10	✓
4"	8	3.01	E	24.08	.20	E	1.60	✓
GLUE	2 qts.	6.00	E	12.00			1.00	✓
P1000 UNISTRUT	20	182.50	C	36.50 ✓	4.50	C	.90	✓
P1490 CLAMPS	15	81.90	C	12.29 ✓	.10	E	1.50	✓
P1431 CLAMPS	8	87.60	C	7.01 ✓	.10	E	.80	✓
Total				331.27 ✓			26.45 ✓

Pricing Sheet

Job: _____   Estimate No.: *M351*

Work: _____   Sheet: *5* of *8*

Estimator: *ED*_____ Checker *SUE* ✔   Date: _____

Description	Qty.	Price	Per	Extension	Hours	Per	Extension
1/2" RIGID CONDUIT	190	37.80	C	71.82 ✔	3.75	C	7.13 ✔
3/4"	10	45.26	C	4.53 ✔	4.00	C	.40 ✔
1"	10	64.08	C	6.41 ✔	4.50	C	.45 ✔
1/2" TERMINATION	83	17.85	C	14.82 ✔	.05	E	4.15 ✔
3/4"	5	27.70	C	1.39 ✔	.06	E	.30 ✔
1"	2	47.25	C	.95 ✔	.08	E	.16 ✔
2" BASE SPACERS	14	35.97	C	5.04 ✔	.05	E	.70 ✔
4" " "	30	49.49	C	14.85 ✔	.05	E	1.50 ✔
TIE WIRE	1 ROLL	3.75	E	3.75 ✔	—	E	1.00 ✔
3/0 EXT. RINGS	2	94.55	C	1.89 ✔	.15	E	.30 ✔
4/S BOXES	188	85.80	C	161.30 ✔	.20	E	37.60 ✔
4 11/16 "	13	156.20	C	20.31 ✔	.30	E	3.90 ✔
4/S P RINGS 5/8"	20	48.70	C	9.74 ✔	.05	E	1.00 ✔
SW. 1 GANG 5/8"	73	50.55	C	36.90 ✔	.05	E	3.65 ✔
SW. 1 GANG 1 1/4"	50	71.15	C	35.58 ✔	.05	E	2.50 ✔
SW. 2 GANG 5/8"	8	78.10	C	6.25 ✔	.06	E	.48 ✔
BLANK	37	33.00	C	12.21 ✔	.05	E	1.85 ✔
4 11/16 SW. RING 5/8"	6	97.75	C	5.87 ✔	.06	E	.36 ✔
SW. RING 1 1/4"	3	95.70	C	2.87 ✔	.06	E	.18 ✔
BLANK	4	54.95	C	2.20 ✔	.06	E	.24 ✔
FS 1	2	4.22	E	8.44 ✔	.30	E	.60 ✔
FSC - 1	2	4.66	E	9.32 ✔	.35	E	.70 ✔
FSS - 1	3	4.66	E	13.98 ✔	35	E	1.05 ✔
BLOCKING/BACKING	201	.70	E	40.20 ✔	.06	E	12.06 ✔
Total				490.62 ✔			82.26 ✔

Pricing Sheet

Job: _____

Work: _____

Estimator: ___ED___ Checker ___SUE ✓___

Estimate No.: *M351*

Sheet: _6_ of _8_

Date: _____

Description	Qty.	Price	Per	Extension	Hours	Per	Extension
#12 THHN	8700	39.80	m	346.26 ✓	6.50	m	56.55 ✓
10	500	60.69	m	30.35 ✓	7.50	m	3.75 ✓
8	160	117.92	m	18.87 ✓	9.50	m	1.52 ✓
6 THW	170	154.80	m	26.32 ✓	11.00	m	1.87 ✓
4	480	234.50	m	112.56 ✓	13.00	m	6.24 ✓
2	360	360.00	m	129.60 ✓	15.00	m	5.40 ✓
1	80	489.20	m	39.14 ✓	16.00	m	1.28 ✓
1/0	360	601.80	m	216.65 ✓	18.00	m	6.48 ✓
#2 BC	50	113.94	m	5.70 ✓	14.00	m	.70 ✓
3/0 "	120	916.16	m	110.01 ✓	18.00	m	2.16 ✓
PULL STRING 200 lb.	880			4.00 ✓			4.00 ✓
MUSIC / PAGING				BY OTHERS			
(S)	19				.25		4.75 ✓
[VC]	7						BY OTHERS
[FM]	2						
[mic]	1						
[S]	4				.30	E	1.20 ✓
			Total	1039.46 ✓			95.90 ✓

Pricing Sheet

Job: _____ Estimate No.: *m351*

Work: _____ Sheet: *7* of *8*

Estimator: *ED*   Checker *SUE* ✓   Date: _____

Description	Qty.	Price	Per	Extension	Hours	Per	Extension
⌀ 1221-I	5	906.50	C	45.33 ✓	.30	E	1.50 ✓
SS	6	1069.25	C	64.16 ✓	.60	E	3.60 ✓
⊖ 5242-I	77	182.35	C	140.41 ✓	.30	E	23.10 ✓
⊖ WP	1	—	E	11.96	—		.35
⊕	2	372.75	C	7.46 ✓	.60	E	1.20 ✓
⊖ GFI SIR-15-F-IV	2	3132.50	C	62.65 ✓	.30	E	.60 ✓
⊖ GFI WP	8	3568.60	C	285.49 ✓	.35	E	2.80 ✓
△ 20	2	310.40	C	6.21 ✓	.30	E	.60 ✓
△ 30	1	—	E	5.28	—	E	.35
△ 50	1	—	E	5.28	—	E	.40
Ⓛ	5	186.55	C	9.33 ✓	.30	E	1.50 ✓
PLUGMOLD 20G12 1 CIR	36'	—	LOT	65.81	—	LOT	4.50
20GA12 2 CIR	48'	—	LOT	103.55	—	LOT	6.00
ADAPTER 2051-H	4	216.59	C	8.66 ✓	.20	E	.80 ✓
END CAP 2010-B	8	18.75	C	1.50 ✓	.10	E	.80 ✓
ELBOW INTERNAL 2017-TC	1	—	E	.34	—	E	.10
⊙	4	34.32	E	137.28 ✓	1.00	E	4.00 ✓
◓	4	34.75	E	139.00 ✓	.80	E	3.20 ✓
◀	19	30.80	C	5.85 ✓	.10	E	1.90 ✓
ST	1	—	E	14.34	—	E	.50
ST WP	2	25.53	E	51.06 ✓	.60	E	1.20 ✓
⊡ 30/3	1	—	E	35.00	—	E	.30
⊡ 30/3 W.P.	3	63.00	E	189.00 ✓	.30		.90 ✓
⊡ 60/3 W.P.	1	—	E	113.50		E	.75
40A FUSES	3	59.00	C	1.77 ✓	.05	E	.15 ✓
T.S. 2P 7202 ZL	2	181.02	E	362.04 ✓	1.00	E	2.00 ✓
⊠ SIZE O 30/3	2	206.39	E	412.78 ✓	1.00	E	2.00 ✓
⊡ 12V	2	1.13	E	2.26 ✓	.20	E	.40 ✓
			Total	2287.30 ✓			65.50 ✓

Pricing Sheet

Job: _____ Estimate No.: *M351*
Work: _____ Sheet: *8* of *8*
Estimator: _____*ED*_____ Checker _*SUE* ✓__ Date: _____

Description	Qty.	Price	Per	Extension	Hours	Per	Extension
⊤ 12V	1	—	E	5.20	—	E	.30
▢ 12V	1	—	E	11.00	—	E	.20
◉ SHUNT TRIP	1	—	E	22.00	—	E	.50
⊠ OH DOOR CONTROLLER	1	—	E	BY OTHERS	—	E	1.50
◉ " " PB STA.	1	—	E	"	—	E	.50
HOOKUP A/C UNIT	1	—	E	15.00	—	E	2.50
EF 1/4 HP	2	5.00	E	10.00✓	.50	E	1.00✓
EF 1 HP	3	5.00	E	15.00✓	.60	E	1.80✓
COMPRESSOR	1	—	E	5.00	—	E	1.00
SUPPLY FAN	2	5.00	E	10.00✓	.75	E	1.50✓
CIR. PUMP	1	—	E	5.00	—	E	.50
WTR HTR	1	—	E	5.00	—	E	.50
O.H. DOOR	1	—	E	10.00	—	E	1.50
4'X8'X 3/4" PLY BACKBOARD	1	—	E	18.00	—	E	1.00
TERM. CAB 12"X18"X4"	2	32.45	E	64.90✓	1.50	E	3.00✓
MARKERS	6	5.00	E	30.00✓	1.00	E	6.00✓
1/2" ROOF FLASHING	5	1.25	E	6.25✓	.30	E	1.50✓
3/4 " "	1	—	E	1.30	—	E	.30
PITCH POCKET 6"X10"X4"	1	—	E	10.00	—	E	1.00
PAD MT. SLAB	1	—	E	285.00	—	E	2.50
5/8"X8' GR. ROD	2	10.38	E	20.76✓	.75	E	1.50✓
5/8" GR. CLAMP	2	1.72	E	3.44✓	.25	E	.50✓
#2 LUG	1	—	E	1.35	—	E	.25
TRENCH 10"X24"	125	1.30	FT	162.50✓	.03	E	3.75✓
" 20"X40"	75	1.70	FT	127.50✓	.05	E	3.75✓
CONCRETE 3000 PSI	6	50.00	CY	300.00✓	.50	E	3.00✓
MTG. FRAME-PANEL "D"	1	—	E	30.00	—	E	2.00
			Total	1174.20 ✓			43.35✓

Telephoned Quotations

Job: ADVERTISING COMPANY.
Supplier: ACE ELECTRIC SUPPLY
Person Quoting: JOHN

Estimate No.: M 351
Estimator: ED
Time: 11:45 A.M.
Date: _____

Quantity	Description	Price	Net
1	FIXTURE TYPE A	50.00/E	
46	A1	32.00/E	
4	B	190.00/E	
1	C	46.00/E	
4	D	40.00/E	
4	E	38.00/E	
1	E1	45.00/E	
2	F	50.00/E	
2	G	75.00/E	
2	H	38.00/E	
2	J	25.00/E	
10	K	1100.00/LOT	
4	L	18.00/E	
12	M	3600.00/LOT	
1	N	29.00/E	
		Total	

F.O.B. Jobsite   YES
Tax Included   NO
Installed   NO
Plans & Specs.   YES
Addendums   1
Division   16

Includes:
_____
_____
_____

Excludes: LAMPS
TAX

Telephoned Quotations

Job: ADVERTISING COMPANY   **Estimate No.:** M 351

Supplier: KING INDUSTRIES   **Estimator:** ED

Person Quoting: PEG   **Time:** 10:20 A.M.

       Date: _____

Quantity	Description	Price	Net
1	FIXTURE TYPE A	44.00/E	
46	A1	28.00/E	
4	B	192.00/E	
1	C	45.00/E	
4	D	42.00/E	
4	E	36.00/E	
1	E1	42.00/E	
2	F	55.00/E	
2	G	75.00/E	
2	H	40.00/E	
2	J	28.00/E	
10	K	1050.00/LOT	
4	L	20.00/E	
12	M	3800.00/LOT	
1	N	30.00/E	
		Total	

F.O.B. Jobsite _____ YES _____   **Includes:**

Tax Included _____ NO _____   _____

Installed _____ NO _____   _____

Plans & Specs. _____ YES _____   _____

Addendums _____ 1 _____   **Excludes:**

Division _____ 16 _____   LAMPS

     TAX

Telephoned Quotations

Job: ADVERTISING COMPANY

Supplier: CENTRAL LIGHTING SUPPLY

Person Quoting: RAY

Estimate No.: M 351

Estimator: ED

Time: 1:05 P.M.

Date: _____

Quantity	Description		Price	Net
1	FIXTURE	A	48.00/E	
46		A1	30.00/E	
4		B	190.00/E	
1		C	42.00/E	
4		D	44.00/E	
4		E	39.00/E	
1		E1	44.00/E	
2		F	50.00/E	
2		G	75.00/E	
2		H	38.00/E	
2		J	22.00/E	
10		K	1075.00/LOT	
4		L	25.00/E	
12		M	4000.00/LOT	
1		N	30.00/E	
			Total	

F.O.B. Jobsite YES

Tax Included NO

Installed NO

Plans & Specs. YES

Addendums 1

Division 16

Includes:
_____
_____
_____

Excludes:
LAMPS
TAX

Telephoned Quotations

Job: ADVERTISING COMPANY

Supplier: SUN ELECTRIC

Person Quoting: BRENDA

Estimate No.: M351

Estimator: ED

Time: 11:00 A.M.

Date: _____

Quantity	Description	Price	Net
1	FIXTURE TYPE A	50.00/E	
46	A1	30.00/E	
4	B	195.00/E	
1	C	46.00/E	
4	D	38.00/E	
4	E	42.00/E	
1	E1	48.00/E	
2	F	48.00/E	
2	G	75.00/E	
2	H	44.00/E	
2	J	25.00/E	
10	K	1000.00/LOT	
4	L	20.00/E	
12	M	3800.00/LOT	
1	N	35.00/E	
	PRICES FIRM - 45 DAYS		
		Total	

F.O.B. Jobsite YES

Tax Included NO

Installed NO

Plans & Specs. YES

Addendums 1

Division 16

Includes:

_____

_____

_____

Excludes:

LAMPS

TAX

Telephoned Quotations

Job: _ADVERTISING COMPANY_   Estimate No.: _M351_
Supplier: _WESTSIDE ELEC. & SUPPLY_   Estimator: _ED_
Person Quoting: _AL DUNN_   Time: _11:25 A.M._
                                      Date: _____

Quantity	Description	Price	Net
1	FIXTURE TYPE A	54.00/E	
46	A1	31.90/E	
4	B	193.00/E	
1	C	55.00/E	
4	D	40.00/E	
4	E	35.00/E	
2	E1	40.00/E	
3	F	45.00/E	
2	G	75.00/E	
2	H	40.00/E	
2	J	25.00/E	
10	K	1045.00/E	
5	L	30.00/E	
12	M	3000.00/LOT	
2	N	40.00/E	
	FIRM 60 DAYS		
		Total	

F.O.B. Jobsite   _YES_   Includes:
Tax Included   _NO_   _____
Installed   _NO_   _____
Plans & Specs.   _YES_   _____
Addendums   _1_   Excludes:
Division   _16_   _LAMPS_
                              TAX

Telephoned Quotations

Job: ADVERTISING COMPANY

Supplier: ACE ELEC. SUPPLY

Person Quoting: ART LONG

Estimate No.: M361

Estimator: ED

Time: 12:20 P.M.

Date: _____

Quantity	Description	Price	Net
	MSB 600A BUS		
	1- 70/3 C/B		
	1- 90/3		
	2- 100/3		
	1- 125/3		
	2- 150/3		
	3 SURFACE PANELBOARDS (1 W/SHUNT TRIP)		
	1 R/T PANELBOARD, WITH SHUNT TRIP		

Total 5,200.00/LOT

F.O.B. Jobsite	YES	**Includes:**	
Tax Included	NO		
Installed	NO		
Plans & Specs.	YES		
Addendums	1	**Excludes:**	
Division	16		

Telephoned Quotations

Job: ADVERTISING COMPANY

Supplier: KING INDUSTRIES

Person Quoting: TONY

Estimate No.: M 351

Estimator: ED

Time: 1:15 P.M.

Date: _____

Quantity	Description	Price	Net
	MSB 600A 120/208 V		
	METERING & DISTRIBUTION SECTION		
	U.G. PULL SECTION NEMA 1		
	2 - 225A PANELBOARDS - SURFACE		
	1 - 100A PANEL WITH SHUNT TRIP MAIN		
	1 - 100A " " " " " R/T		
	Total		5600.00/LOT

F.O.B. Jobsite YES

Tax Included NO

Installed NO

Plans & Specs. YES

Addendums 1

Division 16

Includes:
_____
_____

Excludes:
COPPER BUS
_____
_____

Telephoned Quotations

Job: ADVERTISING COMPANY
Supplier: SUN ELECTRIC
Person Quoting: SHORTY

Estimate No.: M351
Estimator: ED
Time: 10:00 A.M.
Date: _____

Quantity	Description	Price	Net
	MSB W/PULL, METER- CT & DISTRIBUTION SECTIONS		
	1 W.P. PANEL W/SHUNT TRIP MAIN		
	1 PANEL W/SHUNT TRIP MAIN		
	2 PANELBOARDS - SURFACE MOUNTED TOP FEED		

Total 6,500.00/LOT

F.O.B. Jobsite YES
Tax Included NO
Installed NO
Plans & Specs. YES
Addendums 1
Division 16

Includes:

Excludes:

Telephoned Quotations

Job: ADVERTISING COMPANY

Supplier: WESTSIDE ELEC. & SUPPLY

Person Quoting: GEO. BROWN

Estimate No.: M351

Estimator: ED

Time: 9:30 A.M.

Date: _____

Quantity	Description	Price	Net
	MAIN SWITCHBOARD		
	PANEL A & B SURFACE MOUNTED		
	PANEL C W/SHUNT TRIP MAIN		
	PANEL D W/SHUNT TRIP MAIN W.P.		

Total 5950.00/LOT

F.O.B. Jobsite	YES	**Includes:**
Tax Included	NO	
Installed	NO	
Plans & Specs.	YES	
Addendums	1	**Excludes:**
Division	16	

Telephoned Quotations

Job: ADVERTISING COMPANY

Supplier: BLUE & GOLD SOUND, INC.

Person Quoting: BUD PEACE

Estimate No.: M351

Estimator: ED

Time: 8:30 A.M.

Date: _____

Quantity	Description	Price	Net
	AMPLIFIER		
19	RECESSED SPEAKERS		
4	SURFACE "		
7	VOL. CONTROL STA.		
2	FM OUTLETS		
1	MIC & OUTLET		
	INSTALLATION OF WIRE IN YOUR CONDUIT, 1 YEAR SERVICE		
	INCLUDES TAX		

Total 2850.00/LOT

		Includes:
F.O.B. Jobsite	YES	
Tax Included	YES	
Installed	YES	
Plans & Specs.	YES	
Addendums	1	**Excludes:**
Division	16	CONDUIT BOXES

Telephoned Quotations

Job: ADVERTISING COMPANY   **Estimate No.:** M351

Supplier: ELECTRONIC DESIGNS, INC.   **Estimator:** ED

Person Quoting: DON BURTON   **Time:** 10:50 A.M.

Date: _____

Quantity	Description	Price	Net
	FURNISH ALL MUSIC/PAGING EQUIP.		
	INSTALLED CONDUIT SYSTEM - BOXES BY OTHERS		

Total 6,500.00/LOT

		Includes:
F.O.B. Jobsite	YES	
Tax Included	YES	
Installed	YES	
Plans & Specs.	YES	
Addendums	1	Excludes:
Division	16	

Telephoned Quotations

Job: __ADVERTISING COMPANY__   Estimate No.: __M 351__
Supplier: __STUDIO ONE__   Estimator: __ED__
Person Quoting: __PAUL SLOAN__   Time: __12:05 P.M.__
    Date: _____

Quantity	Description	Price	Net
	MUSIC & PAGING SYSTEM		
	ALL SPEC'D EQUIP.	3,800.00	
	INSTALLATION	3,000.00	

Total __6,800.00/LOT__

F.O.B. Jobsite __YES__   Includes:
Tax Included __YES__   _____
Installed __YES__   _____
Plans & Specs. __YES__   _____
Addendums __1__   Excludes:
Division __16__   _____
    _____
    _____
    _____

Telephoned Quotations

Job: ADVERTISING COMPANY

Supplier: SOUND ENGINEERING

Person Quoting: Bob Foss

Estimate No.: M351

Estimator: ED

Time: 9:50 A.M.

Date:

Quantity	Description	Price	Net
	MUSIC/PAGING SYSTEM		
	AMP		
	23 SPEAKERS		
	1 VOL. CONTROL UNITS		
	2 FM RECEPTACLES		
	1 MIC OUTLET		
	1 PAGING MIC		

Total 7,000.00/LOT

F.O.B. Jobsite	YES	**Includes:**
Tax Included	YES	
Installed	YES	
Plans & Specs.	YES	
Addendums	1	**Excludes:**
Division	16	

Spread Sheet

Job: ADVERTISING COMPANY   **Estimate No.:** M 351

Material: _____   **Date:** _____

		ACE JOHN		KING PEG		CENTRAL RAY		SUN BRENDA		WESTSIDE AL	
A	1	—	50.00	—	44.00	—	48.00	—	50.00	—	54.00	
A1	46	32.00	1472.00	28.00	1288.00	30.00	1380.00	30.00	1380.00	31.50	1449.00	✓
B	4	190.00	760.00	192.00	768.00	190.00	760.00	195.00	780.00	193.00	772.00	✓
C	1	—	46.00	—	45.00	—	42.00	—	46.00	—	55.00	
D	4	40.00	160.00	42.00	168.00	44.00	176.00	38.00	152.00	40.00	160.00	✓
E	4	38.00	152.00	36.00	144.00	39.00	156.00	42.00	168.00	35.00	140.00	✓
E1	1	—	45.00	—	42.00	—	44.00	—	48.00	—	40.00	
F	2	50.00	100.00	55.00	110.00	50.00	100.00	48.00	96.00	45.00	90.00	✓
G	2	75.00	150.00	75.00	150.00	75.00	150.00	75.00	150.00	75.00	150.00	✓
H	2	38.00	76.00	40.00	80.00	38.00	76.00	44.00	88.00	40.00	80.00	✓
J	2	25.00	50.00	28.00	56.00	22.00	44.00	25.00	50.00	25.00	50.00	✓
K	10	LOT	1100.00	LOT	1050.00	LOT	1075.00	LOT	1000.00	LOT	1045.00	
L	4	18.00	72.00	20.00	80.00	25.00	100.00	20.00	80.00	30.00	120.00	✓
M	12	LOT	3600.00	LOT	3800.00	LOT	4000.00	LOT	3800.00	LOT	3000.00	
N	1	—	29.00	—	30.00	—	30.00	—	35.00	—	40.00	
			7862.00	✓	7855.00		8181.00		7923.00		7245.00	✓

Spread Sheet

Job: _ADVERTISING COMPANY_   **Estimate No.:** _M 351_

Material: _SWITCH GEAR_   **Date:** _____

			ACE ART	KING TONY	SUN SHORTY	WESTSIDE GEO.		
			MSB	✓	✓	✓		
			PANEL-A	✓	✓	✓		
			B	✓	✓	✓		
			C	✓	✓	✓		
			D	✓	✓	✓		
			5,200.00 ✓	5,600.00 ✓	6,500.00 ✓	5,950.00 ✓		

Spread Sheet

Job: ADVERTISING COMPANY   **Estimate No.:** M351

Material: MUSIC & PAGING SYSTEMS   **Date:** _____

			BLUE & GOLD BUD	ELECTRONIC DESIGNS DON	STUDIO ONE PAUL	SOUND ENG. BOB	
		AMPLIFIER	✓	✓	✓		
19	Ⓢ	SPEAKERS	✓	✓	✓		
4	[S]	SPEAKERS	✓	✓	✓		
7	[VC]	VOL. STA.	✓	✓	✓		
2	[FM]	FM OUTLETS	✓	✓	✓		
1	[MIC]	MIC & OUTLET	✓	✓	✓		
		INSTALLED	✓	✓	✓		
	TAX	INCLUDED	✓	✓	✓		
			2,850.00	6,500.00	6,800.00	7,000.00	

Bid Summary

Job: **ADVERTISING COMPANY** Estimate No.: **M 351**
Location: **INDUSTRIAL PARK** Bid Date: _____
Division: **16A ELECTRICAL** Time: **2:00 P.M**
Estimator: **ED** Checker: **SUE ✓**

	Description	Material	Deducts		Labor
1.	SWITCHGEAR	5 000.00	+ 200.00		52.65
2.	FIXTURES	8 278.00	- 1033.00		78.00
	LAMPS	1 273.00	—		9.90
3.	CONDUIT/FITTINGS	1 225.00	—		211.66
4.	" "	331.00	—		26.45
5.	" BOXES	491.00	—		82.26
6.	WIRE	1 039.00	—		95.90
7.	DEVICES	2 287.00	—		65.50
8.	HOOKUP/TRENCH	1 174.00			43.35

Description	Material	Deducts		Labor
Sub Total	21 099.00 ✓	- 833.00 ✓		665.67 ✓
Sales Tax 3 %	633.00	- 25.00 ✓	Supervision 15 %	99.85 ✓
Sub Total	21 732.00 ✓	- 858.00 ✓	Total Hours	765.52 ✓
Labor	8 084.00 ✓		Rate	8.00
Tools	400.00 ✓		Sub Total	6 124.16 ✓
Miscellaneous	211.00		Fringes 8%	489.93 ✓
Permits/Fees	100.00		Taxes 12 %	734.90 ✓
Sub Total	30 527.00 ✓		Sub Total	7 348.99 ✓
Subcontracts ✳	2 850.00		Factor 10%	734.90 ✓
Travel Expense	N/A		Total	8 083.89 ✓
Sub Total	33 377.00 ✓			
Deducts	- 858.00 ✓		Addendums	1
Net Cost	32 519.00 ✓		Alternates	NONE
Overhead 12%	3 902.00 ✓		Job Duration	180 CAL. DAYS
Sub Total	36 421.00 ✓		Penalty	$500.00/DAY
Profit 10 %	3 642.00 ✓		Type Const.	NEW 6,000 #
Contingency FIXTURE 5%	362.00 ✓			
Bond 15.00/M	546.00 ✓			
Selling Price	40 971.00 ✓			

BID PRICE 40,965.00   ✳ SOUND SYSTEM PRICE

Chapter 12

Bidding Mechanics

There are a few general ground rules to estimating that every electrical estimator should understand. Most seasoned estimators know the points outlined in this chapter, and you should too.

Get to Know the General Contractor
Make it a point to visit general contractors occasionally. Get to know them and their chief estimators. Establish good working relationships with them. Show that you're sincerely interested in working with them. They're probably negotiating contracts that you might be able to bid. Every estimator needs to be a little bit of salesman.

Follow up on the jobs you've bid. Study competing prices. General contractors are usually happy to show you the quotations they received on a job you bid. Check back with the contractor after the winning bid has been selected.

Naturally, general contractors prefer subcontractors who do professional quality work and do it on time. They may even give favored subcontractors a last look at the competing bids. This can be helpful in setting your final bid price. But it can also lead to problems. Any contractor that gives you an unfair advantage may be giving others similar treatment. That

leads to distrust. Be cautious about any job offered if you can beat the lowest bid by a few dollars. The bid you have to beat may be a money loser.

Bid Log
Keep a record of all your bids. This record, or log, will be a valuable reference for bidding future jobs. Use a loose-leaf notebook. Write the name of each general contractor you've bid to on separate sheets of notebook paper. Then arrange these sheets in alphabetical order. Make columns in each sheet for the job name, the bid date, your bid price, and the name and price of the electrical contractor awarded the job. (See Figure 12-1.) Record any other information you feel is important.

After a while you'll see certain patterns develop. You may notice that some subs offer better prices to a general contractor who does his job well and finishes on schedule. They know that this makes their work easier. On the other hand, a general contractor who tends to be a poor supervisor, and who constantly runs over budget and behind schedule, will draw fewer and higher bids. Subcontractors know that he will cost them too much time and money.

Contractor Award Record

General Contractor   Associated Contractors, Inc.

Project	Date	Our Price	Awarded To	Price
ADVERTISING CO.	—	40,965.00	US	
MAIN LANES	—	90,850.00	ART'S ELECTRIC	88,888.00
MARK'S DONUTS	—	20,155.00	ART'S ELECTRIC	20,055.00
ST. LUKE'S CHURCH	—	77,750.00	ART'S ELECTRIC	77,500.00

Figure 12-1

Contractor Award Record

General Contractor   Billy Builders

Project	Date	Our Price	Awarded To	Price
SMITH SCHOOL	—	77,870.00	HUMTBLE ELEC.	75,444.00
HELEN'S BAKERY	—	18,800.00	ARROW ELEC.	17,950.00
GINA'S MARKET	—	33,650.00	J-W & SONS	30,000.00

Figure 12-1 (continued)

Contractor Award Record

General Contractor   Custom Construction Company

Project	Date	Our Price	Awarded To	Price
MERCY INST	——	105,555.00	ANDY'S ELEC.	103,000.00
ACE HOTEL	—	95,650.00	ALGARD ELEC.	95,625.00
CITY LIBRARY	——	177,850.00	S+S ELEC.	166,666.00
HALL OFFICES	—	85,500.00	US	—

Figure 12-1 (continued)

Contractor Award Record

General Contractor   Jefferson Construction Works

Project	Date	Our Price	Awarded To	Price
STAR MARKET	—	75,000.00	JENNINGS ELEC.	74,750.00
ATLAS HOSP.	—	168,850.00	JENNINGS ELEC.	161,161.00
HOME PRINTERS	—	70,000.00	JENNINGS ELEC.	65,555.00

Figure 12-1 (continued)

Contractor Award Record

General Contractor   States Best Builders

Project	Date	Our Price	Awarded To	Price
BILL'S GARAGE	——	12,250.00	HUMBLE ELEC.	12,100.00
SOUTH SCHOOL	——	175,000.00	WILCOX ENTERP.	170,750.00

Figure 12-1 (continued)

Contractor Award Record

General Contractor   Younger Brothers & Associates, Inc.

Project	Date	Our Price	Awarded To	Price
BEEF & STEW	—	29,800.00	ART'S ELEC	25,666.00
EAST MALL	—	248,450.00	SPARKY'S ELEC.	258,000.00

Figure 12-1 (continued)

You can use information like this to your advantage. Keeping a bid log will help you identify special relationships and attitudes that affect the way you do business.

Subcontractor Listing

Many bidding documents, require the general contractor to complete a subcontractor form provided by the owner. This form lists the work to be performed and the subcontractor's name, address, and license number. The general contractor must use the subs listed unless a subcontractor can't be bonded or financial difficulties make it unlikely that he'll finish the job. The general contractor is bound by the price of his bid to the owner. But the general contractor is still responsible for the work of each sub.

Some bidding documents call for the subcontractor listing at the time the contract is awarded. This is usually a disadvantage to the subcontractors because the general contractor can shop the job for weeks before the contract is awarded. Your bid may have been the best at bid time and probably helped the general contractor win the contract. Still, you might not get the job.

Fortunately, few general contractors shop bids. They know that other jobs will come along and they'll need your help to get that work.

Scope of Bid

Develop a standard form that can be sent to the general contractors to advise them of what you intend to bid. The form should indicate the project, the bid date, the sections you intend to bid and the items to be added or excluded. A simple postcard or form can do the job. (See Figure 12-2.) Send this information to the general contractor as early as possible.

Determine which areas of your work are best handled by the general contractor. The more work you can divert to other trades, the lower your overall bid price will be. Light pole bases and concrete equipment pads can be installed by the general. Work this out with the contractor well before the bid date.

Most general contractors use a standard percentage to calculate the cost of temporary electrical service. Usually it's better if you exclude the cost of temporary service from your bid. But some general contractors want that cost included. The size and voltage of the service, the amount of lighting and the locations for power outlets usually determine the actual costs. Some contractors base the cost of temporary service on some other criteria. Check with the contractor on each project. And don't make a practice of automatically including temporary service on your bids.

Telephone Bidding

Telephone the general contractor to quote the specific electrical work you're bidding. Always include the following information in your telephone bid:

- The name of your company.
- Your name. Spell it so there's no doubt. And write down the name of the person you're speaking with.
- Your telephone number.
- The section you're bidding.
- Whether or not the bid is in accordance with the plans and specifications.
- Whether or not the bid includes taxes, fees and installation.
- Any additional items that are included.
- All exclusions from the section you are quoting.
- The bid price. Make a practice of correctly stating the full dollar amount; avoid shortcuts or slang.

Ask the other party to repeat the entire bid and correct any errors he may have made. Let him know that you are available throughout the bid period should any clarifications be necessary.

Follow up with a written quotation listing all of the above items. Be sure to keep track of the contractors you're bidding to, the amount bid, and who you talked to.

A telephone bid is binding. Your company must be willing to accept a contract for the work stated in the bid at the bid price. That's one good reason for having the bid read back to you over the phone. It's also good practice to have the estimator make all telephone bids. It helps eliminate confusion, reduce errors and clear up any questions or problems which may arise.

Some general contractors may suggest that your price is too low or too high compared to other bids. Be wary of this tactic. A quick check or review of the end sheet should justify the accuracy and completeness of your bid. This isn't the time to go all the way through the estimate looking for a big mistake. If

Bidding Scope Form

Project Name: Advertising Company   **Bid Date:** 7/20/82
Location: Industrial Park
Section Bidding: 16A Electrical   **Estimator:** Ed

Scope: Electrical Work in accordance with the plans and specifications.

Music and Paging System.

Hookup Mechanical Equipment.

Includes: Blocking for electrical lighting fixtures and for electrical equipment.

Excludes: Temporary Electrical Service and Temporary Electrical Wiring.

Figure 12-2

you're not sure of your price, withdraw the bid.

Some general contractors may ask you to lower your cost to the next lower cost. Shady operators enjoy working one sub against another. Don't take a job just to keep busy. Your company should earn a profit on all work. The margin for profit and the risks involved are much too great to gamble on bidding at or below cost.

When you're sure that a general contractor will take your bid in strict confidence, bid the job as early as possible. Contractors appreciate early bids. They'll be receiving many bids from other trades during the last hours of the bid day. An early, reliable bid can be a big help to the contractor.

Hold back bids to contractors you feel will probably peddle your price to your competitors. Submit those bids as late as possible. This limits their ability to shop your bid.

On larger projects, the general contractor may ask if your company can bond the job. Usually the performance bond will be set at 100 percent of the bid price. If your company can't bond the job, say so. The contractor will know if your firm can handle the job without bond. If you say that bonding can be furnished when it can't, your company may be liable for the difference in the bid price to the next responsible subbidder. You've also lost the trust of the general contractor.

Complimentary Bid

A competitor may ask you to make a *complimentary* or *courtesy bid* for him. Perhaps he doesn't have time to make the take-off for an estimate and one of his best general contractor customers is depending on him for a specific job. Your competitor doesn't want to know how much you are going to bid, he just wants a higher price to give to his customer. He also wants to keep good relations with his customer. Be careful when giving complimentary bids, and don't encourage their practice. You might be limiting competition, which may be illegal.

Collusion

Collusion is a secret agreement between two or more persons for a deceitful, dishonest or fraudulent purpose. Don't discuss pricing, scope or bidding tactics with any of your competitors before bid time. Any discussion about the bid can be classified as collusive. If a competitor calls to discuss an estimate, don't do it. It is illegal. Many contractors and electrical manufacturers have paid large penalties and fines for fixing prices among themselves.

Chapter 13

Submittals

Most construction contracts require the contractor to provide submittal data for material and equipment purchased for the project. Either the general or special conditions of the contract specify the method for preparing and presenting submittal data. Each subsection of the specifications lists the detailed requirements for that section. A subsection may require certain shop drawings, specific catalog sheets and a general material list.

Shop Drawings
Shop drawings are usually required for special equipment such as switchboards, motor control centers and substations. The manufacturer prepares the drawings prior to fabrication. The drawings must reflect all of the requirements of the project's plans and specifications. Usually a single switchboard consists of several drawings. One drawing is prepared for the construction of the housing, another for the schematic wiring. Details or schedules are listed for circuit breakers and nameplate data. Additional drawings may be required for other design criteria.

Catalog Sheets
Catalog sheets are used for submittal data on many items. The catalog sheets for a certain lighting fixture should include the lens type, photometrics, voltage and accessories. Each sheet should be marked to indicate the fixture type shown on the project's lighting fixture schedule. Each detail should be marked also.

Catalog sheets for the magnetic starters, for example, can be copied from the manufacturer's catalog. The specific magnetic starter, voltage, size and enclosure should be plainly marked. (See page 195.)

Material List
The material list should include the type, size, manufacturer and catalog number of the material specified. It should include each item of equipment shown on the shop drawings or covered by a catalog sheet. The list should be as complete as possible; it may serve as an index to the submittal package. (See page 171.)

Preparing the Submittal
After the contract is awarded, the first step in preparing the submittal is to send a purchase order for the material or equipment to the supplier who quoted the best price at bid time. Make it a condition of the purchase that all items comply with the plans and specifications.

Note on the purchase order that the items will be delivered after approval of the submittal. List the number of shop drawings and catalog sheets that will be required for submittal.

Most suppliers will mark the catalog sheets so that all items supplied will be in order. Some furnish an index to the lighting fixtures for easier review.

After the purchase orders are sent to the suppliers, start to assemble the material list.

Prepare a cover sheet for the submittal package. List the name and location of the project at the top. On the right-hand side list the names of the architect, electrical engineering firm, general contractor and your company.

Head the second page "Electrical Material List." Turn the paper sideways and head the columns to identify item, material, size, type, and manufacturer. Allow at least one inch for binding.

List as many manufacturers as you can for the basic material items. Listing only one manufacturer will limit your field. Most suppliers carry more than one manufacturer's products, so check for a list of the manufacturers they most commonly stock.

The material list shows the architect or his consulting engineer the quality of the material being offered. Most engineers don't demand specific material manufacturers for minor fittings. But they do care about the special items such as switchgear and lighting fixtures.

Most contracts require submittal of a minimum number of brochures and shop drawings for approval. It's best to send a few extras. If you send the required number, there may not be enough to go around. Plan to send at least ten sets, and request that at least three be returned. One set will be needed for your files, one for returning to the supplier and one for installing the work.

Most manufacturers won't start fabrication of the equipment until the shop drawings are returned and marked "Approved" or "Approved as Noted." You'll need the two other sets to do the job.

Both the architect and the engineer get a couple of sets. Other designers may need a copy, and the general contractor will also need a couple of sets. If there's an on-site inspector for the job, he'll also need a set.

A complete submittal package is given at the end of this chapter.

Submittal Review

Deliver the submittal package to the general contractor along with a transmittal letter identifying the project. The letter should also request that at least three sets of brochures and shop drawings be returned so that fabrication can be scheduled.

The general contractor forwards the submittal to the architect, recommending approval of the package. He also requests that at least five sets of brochures and shop drawings be marked "Approved" and returned as early as possible so that the equipment and material can be purchased.

The architect forwards the submittal to the consulting engineer for review. If all is in compliance with the plans and specifications, the engineer stamps the package "Approved" and returns it to the architect, who in turn sends it to the general contractor.

If the engineer finds that the submittal does not comply with the specifications, it is either rejected or marked "Approved as Noted." If rejected, a new submittal must be prepared. If marked "Approved as Noted," it can be sent to the supplier with instructions to make the corrections before fabrication.

Use extra care when preparing submittals for projects that have a short construction period. A resubmittal will take too much time and may cause a delay in obtaining the right material.

Return of Submittals

When submittals are returned from the general contractor, remove them from the binding and separate them. Return the lighting fixture catalog sheets to your supplier as soon as possible. Be sure to include a transmittal letter. Do the same with the panelboard shop drawings.

Prepare a job folder for field work and attach a copy of the submittal data. The field crew will need the information early so that the rough-in work can proceed smoothly.

Rejected Submittal

Sometimes there are enough errors in the submittal to warrant rejection of a certain item. Yet the engineer may approve all of the other parts. Usually the rejected item can be singled out and corrected with a simple resubmittal. Do that right away to avoid delaying the project.

There are times when the supplier can't furnish all of the lighting fixture types specified. A lighting fixture resembling a specified item may be submitted instead. Usually the engineer will understand the problem and may approve the alternate fixture. If your supplier tries to substitute all lower quality products, the submittal will probably be rejected.

When the supplier quotes a job, he's expected to furnish equipment that complies with the plans and specifications. Additional costs cannot be invoiced for correcting the submittal data.

If the material and equipment submittal complies with the plans and specifications and the engineer rejects it in favor of something else or for some new development, a change order should be negotiated through the general contractor. If the architect or general contractor insists that you proceed with the changes without any modification in the contract, do so only under protest. If you hold up the project, the owner may assess liquidated damages against you. Sometimes these costs can be more than the cost difference for the new material or equipment.

If you find out that there will be a delivery problem, immediately notify the general contractor so that he can inform the architect. The architect may want to substitute something that is available and that will fit into the project schedule. Be sure you're right before claiming something is unavailable.

Advertising Company

Industrial Park

Raymond C. Alcaraz & Assoc.
Architect

Algard Engineering, Inc.
Electrical Engineer

Associated Contractors, Inc.
General Contractors

Your Company's Name
Electrical Contractor

Submittal Package — Cover Sheet
Figure 13-1

ELECTRICAL MATERIAL LIST

ITEM	MATERIAL	SIZE	TYPE	MANUFACTURER
1.2	Rigid Conduit	1/2" to 2"	Galv.	Jones Laughlin, Triangle, Rome
1.2	Couplings	1/2" to 2"	Galv.	Jones Laughlin, Triangle, Rome
1.2	Locknuts	1/2" to 2"	Galv.	Bridgeport, Efcor, Raco, T&B, Appleton
1.2	Bushings	1/2" to 2"	Plastic	Bridgeport, Efcor, Raco, T&B, Appleton
1.3	EMT Conduit	1/2" to 2"	Galv.	Jones Laughlin, Triangle, Rome, Western Tube
1.3	Couplings, W.T.	1/2" to 2"	Galv.	Bridgeport, Efcor, Raco, T&B, Appleton
1.3	Connectors	1/2" to 2"	Galv.	Bridgeport, Efcor, Raco, T&B, Appleton
1.4	Flexible Conduit	1/2" to 1"	Galv.	Triangle, American Flex. Conduit
1.4	Connectors, Jake	1/2" to 1"	Galv.	Bridgeport, Efcor, Raco, T&B, Appleton
1.5	Oil Tight Flex.	1/2" to 1"	Sealtite	Anaconda, Liquid-Tite
1.5	Connectors	1/2" to 1"	Galv.	Bridgeport, Efcor, Raco, T&B, Appleton
1.6	PVC Conduit	1/2" to 4"	Schedule 40	Carlon, Western Plastics
1.6	Couplings	1/2" to 4"	Schedule 40	Carlon, Western Plastics
1.6	Adapters	1/2" to 4"	Schedule 40	Carlon, Western Plastics
1.6	Elbows	1/2" to 4"	Schedule 40	Carlon, Western Plastics
1.7	Wire, Cu	#12 to #8	THHN, Sol/Str.	Carlon, Gen Cable, Hatfield, Rome, Paranite
1.7	Wire, Cu	#6 to #1/0	THW, Stranded	Gen Cable, Hatfield, Rome, Paranite, G.E.
1.8	Lighting Fixtures			See Catalog Cuts
1.8	Lamps	As Specified	W.W.	General Electric, Westinghouse, Sylvania
1.8	Ballasts		See Catalog Cuts	
1.9	Outlet Boxes	As Required	Galv.	Appleton, Bowers, Raco

Submittal Package — Electrical Material List
Figure 13-1 (continued)

ITEM	MATERIAL	SIZE	TYPE	MANUFACTURER
1.9	Plaster Rings	As Required	Galv.	Appleton, Bowers, Raco
1.9	Switch Rings	As Required	Galv.	Appleton, Bowers, Raco
1.9	Cast Metal Boxes	As Required	Cast	Appleton, Bell, Crouse-Hinds
1.10	Floor Boxes	As Required	Adjustable	Hubbell
1.10	Carpet Flanges	5 to 1/4"	Plastic	Hubbell
1.11	S.P. Switches	15A, 277V	1221-I	Leviton, Hubbell
1.12	Duplex Outlets	15A, 125V	5242-I	Leviton, Hubbell, P&S
1.12	W.P. Covers	Duplex	Fiberglass	Leviton, Hubbell, P&S
1.12	G.F.I. Duplex	15A, 125V	SIR-15-F-IV	Slater
1.12	Special Outlets	30A	2610	Hubbell
1.12	Special Outlets	50A	9360	Hubbell
1.13	Cover Outlets	As Required	87000	Leviton
1.14	Plugmold	1 & 2 Circuits	20G12/20GA12	Wiremold
1.14	Fittings	As Required	Plugmold	Wiremold
1.15	Main Switchboard		See Shop Drawings	
1.16	Panelboards		See Shop Drawings	
1.17	Terminal Cabinets		See Shop Drawings	
1.19	Time Switches	40A, 2P	7202ZL	Tork
1.20	Safety Switches		See Catalog Cuts	
1.20	Fuses	As Required	Dual Element	Buss, Econ, Pierce, Monarch, Shawmut
1.21	Push Buttons	120V, 1-P	1251 MC	Hubbell
1.21	Push Buttons	L.V.	1750	Hubbell
1.22	Grounding			

Submittal Package — Electrical Material List
Figure 13-1 (continued)

ITEM	MATERIAL	SIZE	TYPE	MANUFACTURER
1.22	Grounding	3/0	B.C.	Anaconda, Gen Cable
1.22	Ground Clamp	2"	Water Pipe	Raco, T&B, Appleton
1.23	Ground Test	500 VDC	Resistance	Biddle Instruments

Submittal Package — Electrical Material List
Figure 13-1 (continued)

ADVERTISING COMPANY

Lighting Fixtures

TYPE A	Lithonia	2GP440RN A12 ES
TYPE A1	Lithonia	2GP240RN A12 ES
TYPE B	Lithonia	4SG640RN A12 ES
TYPE C	Lithonia	SC240 ES
TYPE D	Prescolite	1015F-3
TYPE E	Prescolite	73222
TYPE E1	Prescolite	73322
TYPE F	Marco	NKP22
TYPE G	Lithonia	WA 296A ES
TYPE H	Marco	B-113
TYPE J	Marco	M1-11
TYPE K	Holophane	2038-120-PHCA-UPH-35-120
TYPE L	Appleton	G-50214
TYPE M	Westinghouse	FR-GSNG-W76A
TYPE N	Lithonia	UN 240 ES

Westside Electrical Supply Co.

Submittal Package—Lighting Fixture Index
Figure 13-2

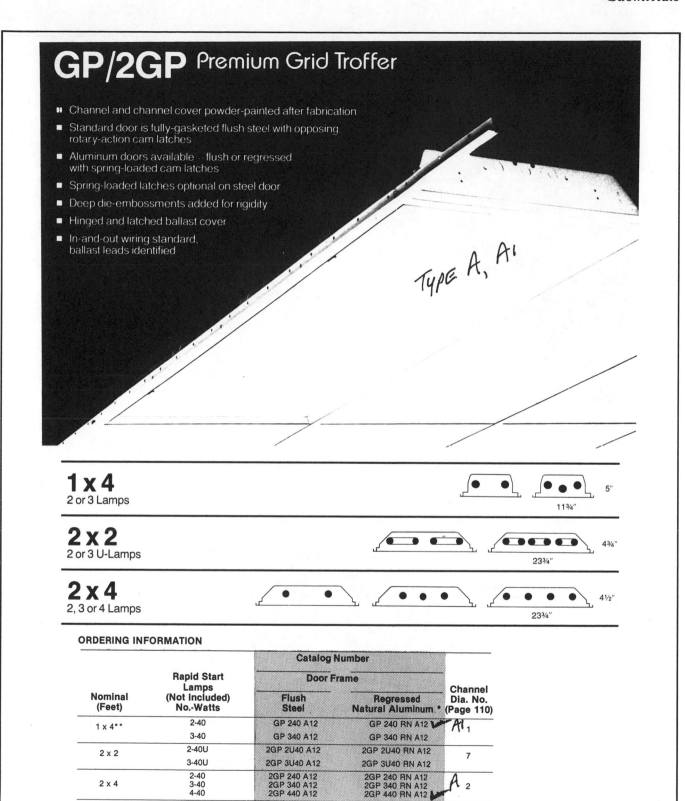

GP/2GP Premium Grid Troffer

- Channel and channel cover powder-painted after fabrication
- Standard door is fully-gasketed flush steel with opposing, rotary-action cam latches
- Aluminum doors available — flush or regressed with spring-loaded cam latches
- Spring-loaded latches optional on steel door
- Deep die-embossments added for rigidity
- Hinged and latched ballast cover
- In-and-out wiring standard, ballast leads identified

TYPE A, A1

1 x 4
2 or 3 Lamps

5"
11¾"

2 x 2
2 or 3 U-Lamps

4¾"
23¾"

2 x 4
2, 3 or 4 Lamps

4½"
23¾"

ORDERING INFORMATION

Nominal (Feet)	Rapid Start Lamps (Not Included) No.-Watts	Catalog Number		Channel Dia. No. (Page 110)
		Door Frame		
		Flush Steel	Regressed Natural Aluminum *	
1 x 4**	2-40	GP 240 A12	GP 240 RN A12	*A1* 1
	3-40	GP 340 A12	GP 340 RN A12	
2 x 2	2-40U	2GP 2U40 A12	2GP 2U40 RN A12	7
	3-40U	2GP 3U40 A12	2GP 3U40 RN A12	
2 x 4	2-40	2GP 240 A12	2GP 240 RN A12	*A* 2
	3-40	2GP 340 A12	2GP 340 RN A12	
	4-40	2GP 440 A12	2GP 440 RN A12	

A12 indicates # 12 pattern acrylic prismatic lens. Optional shielding available.
*Flush aluminum door and optional finishes available.
**1' x 4' models not equipped with wiring access plate.

Submittal Package—Lighting Fixtures
Figure 13-3

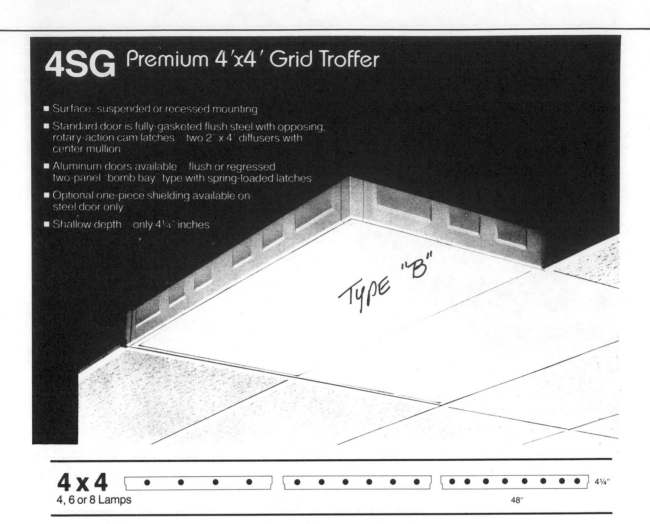

4SG Premium 4'x4' Grid Troffer

- Surface, suspended or recessed mounting
- Standard door is fully-gasketed flush steel with opposing, rotary-action cam latches two 2' x 4' diffusers with center mullion
- Aluminum doors available flush or regressed two-panel "bomb bay" type with spring-loaded latches
- Optional one-piece shielding available on steel door only
- Shallow depth only 4¼" inches

TYPE "B"

4 x 4
4, 6 or 8 Lamps

4¼"

48"

ORDERING INFORMATION

Nominal (Feet)	Rapid Start Lamps (Not Included) No.-Watts	Catalog Number — Door Frame		Channel Dia. No. (Page 110)
		Flush Steel	Regressed Natural Aluminum *	
4 x 4	4-40	4SG 440 A12	4SG 440 RN A12 BD	11
	6-40	4SG 640 A12	4SG 640 RN A12 BD	
	8-40	4SG 840 A12	4SG 840 RN A12 BD	

B

A12 indicates # 12 pattern acrylic prismatic lens. (two 2' x 4' panels). Optional one-piece shielding available (steel door only).
*Flush aluminum door and optional finishes available.
Aluminum doors supplied in two-panel "bomb bay" configuration only.

Submittal Package—Lighting Fixtures
Figure 13-3 (continued)

SC Low-Brightness Wraparound

- Flat-bottom acrylic prismatic diffuser
- Linear side prisms control brightness, pyramidal bottom prisms minimize lamp image
- Continuous, interlocking diffuser support prevents accidental opening, simplifies cleaning and service
- For surface or stem mounting, unit or row installation shift-lock couplers for row mounting without tools
- Order end plates separately   one pair per fixture or row
- UL listed for mounting on low-density ceilings

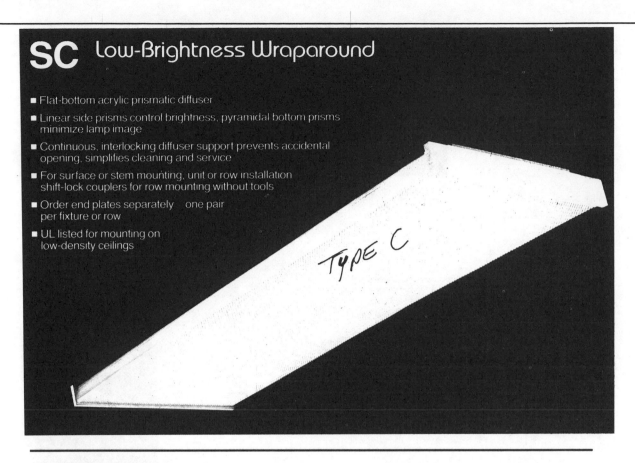

TYPE C

NARROW BODY 14⅝″
4-Foot Length, 2 or 4 Lamps
8-Foot Length, 4 or 8 Lamps

3″   14⅝″

WIDE BODY 17¾″
4-Foot Length, 2 or 4 Lamps
8-Foot Length, 4 or 8 Lamps

3″   17¾″

ORDERING INFORMATION

Nominal Length	Rapid Start Lamps (Not Included) No.-Watts	Catalog Number		Channel Dia. No. (Page 110)
		14⅝″ Wide	17¾″ Wide	
1-Foot	2-40	SC 240 A	2SG 240 A	14
1-Foot	4-40	SCN 440 A	SC 440 A	14
8-Foot	4-40	8TSC 240 A	8T2SC 240 A	21
8-Foot	8-40	8TSCN 440 A	8TSC 440 A	21

Injection-Molded End Plates
Fixtures supplied without end plates. Order separately, one pair per fixture or row.

Catalog Number			Description
White	Black	Bronze	
SC2WH	SC2BL	SC2BR	Decorative end plates for **narrow body** — 1 pair
SC4WH	SC4BL	SC4BR	Decorative end plates for **wide body** — 1 pair

---

Submittal Package—Lighting Fixtures
Figure 13-3 (continued)

1015F-1 Flat Alba Glass

1015F-3 Drop Opal Glass

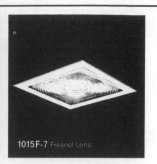
1015F-7 Fresnel Lens

1015 Series 150W, A-19
RECESSED SQUARES.
Specification grade quality Prewired Recessed Housings. Choice of Diecast Relamp-a-lite* or torsion spring retained trims. Matte white finish. U.L. listed for thru branch circuit wiring for 6-60° or 8-75° C. conductors. Suitable for damp locations. Adjustable socket holder. Slip Strap Bar Hangers included for up to 24" O.C. joists.

DIECAST RELAMP-A-LITE TRIM*
1015HF-1 with Flat Alba Glass
1015HF-3 with Drop Opal Glass
1015HF-7 with Fresnel Lens

DIECAST ECONOMY TRIM
1015HS-1 with Flat Alba Glass
1015HS-3 with Drop Opal Glass
1015HS-7 with Fresnel Lens

*Relamp-a-lite trims allow glass to be removed through swing-way hinges for relamping without disturbing trim or ceiling finish.

Accessories:
Packet 261 (Converts Bar Hangers to support for 2' grid suspended ceilings)

TYPE D

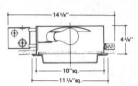

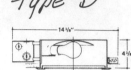

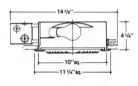

1015S-1 Flat Alba Glass

1015S-3 Drop Opal Glass

1015S-7 Fresnel Lens

... general illumination, soffit

prescolite **13**

Submittal Package—Lighting Fixtures
Figure 13-3 (continued)

FACEPLATES
FOR PROFILE EXITS

Numbers shown are Faceplates only—combine with housing number for complete unit. Three styles of Faceplates available: Stencil without arrows, Stencil with snap-in Universal Arrows, and Open Frame. Standard 6″ high x ¾″ stroke letters to meet O.S.H.A. and local building codes. Special lettering on request* (see pg. **7**). Unbreakable Polymer Diffusers for minimum maintenance. No visible fasteners, unique torsion bars provide positive spring-tension closure and hinged opening for rapid relamping. Knockouts permit tamperproof screws. Matte Oyster finish standard, other finishes available.

HOW TO SPECIFY COMPLETE UNITS

Select required housing mounting condition (Green numbers on pg. **4**). Combine with faceplate style desired selected from illustrated group (Black numbers this pg.)

I.E. Single Face Universal Mounting—**711**
Open Frame Style
White letters, Red background—**33**

71133

Housings or faceplates can be specified separately by individual numbers indicated —i.e. 73200 Housing only, 70021 Faceplate only.

Type E & E1 (handwritten)

STENCIL STYLE

EXIT
70011

EXIT
70012

**STENCIL STYLE
w/UNIVERSAL ARROWS**

E▸XIT◂
70021

E▸XIT◂
70022

OPEN FRAME STYLE

EXIT
70031

EXIT
70032

EXIT
70033

EXIT
70034

E▸XIT◂
70041

E▸XIT◂
70042

E▸XIT◂
70043

E▸XIT◂
70044

CAT. NUMBER TO ORDER FACEPLATES w/ARROW LEFT and w/ARROW RIGHT FOR USE WITH TWIN FACE UNIVERSAL MOUNT and TWIN FACE TRIANGULAR MOUNT

70071

70072

70073

70074

EXIT▸
70051

EXIT◂
70052

EXIT▸
70053

EXIT◂
70054

◂EXIT
70061

◂EXIT▸
70062

◂EXIT
70063

◂EXIT▸
70064

PRESCOLITE **5**

**Submittal Package—Lighting Fixtures
Figure 13-3 (continued)**

MEDIUM REFRACTORS 150W

Medium size non-destructible poly-carbonate injection molded prismatic refractor. Cast conduit pan has four ½" I.P.S. openings and plugs. Steel pan has knock-outs.

Type F

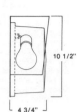

10 1/2"

4 3/4"

150W A21	prismatic regular	prismatic max-secure
Concealed socket pan	NP21	NP22
w/cast conduit pan	NP21C	NP22C
w/steel conduit pan	NP21CP	NP22CP
w/shroud roof	NKP21	NKP22 *F*
Wall bracket	NBP21	NBP22

4¾"

9⅞"

5 3/4"

9 7/8"

6 7/8"

12"

NKP21

NP21C

NBP21

Specify finish [except NP21 & NP22]
P white paint
S satin aluminum
  (available cast pan & wall bracket only)
K black paint
M medium bronze

U.L. Listed for damp locations
I.B.E.W Union made

Submittal Package—Lighting Fixtures
Figure 13-3 (continued)

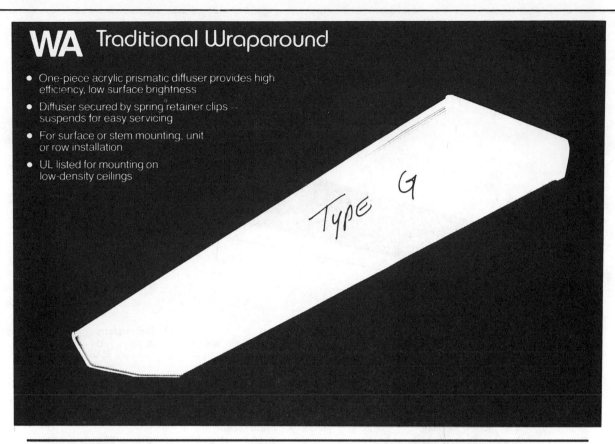

WA Traditional Wraparound

- One-piece acrylic prismatic diffuser provides high efficiency, low surface brightness
- Diffuser secured by spring retainer clips -- suspends for easy servicing
- For surface or stem mounting, unit or row installation
- UL listed for mounting on low-density ceilings

Type G

NARROW BODY 12"
4-Foot Length, 2 Lamps
8-Foot Length, 2 or 4 Lamps

3½"

12"

WIDE BODY 18"
4-Foot Length, 2 or 4 Lamps
8-Foot Length, 4 or 8 Lamps

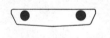 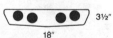

3½"

18"

ORDERING INFORMATION

Nominal Length	Lamps (Not Included) No.-Watts	Type	Catalog Number 12" Wide	18" Wide	Channel Dia. No. (Page 110)
4-Foot	2-40	Rapid Start	WA 240 A	2WA 240 A	15
4-Foot	4-40	Rapid Start	—	WA 440 A	15
8-Foot	4-40	Rapid Start	8TWA 240 A	8T2WA 240 A	22
8-Foot	8-40	Rapid Start	—	8TWA 440 A	22
8-Foot	2-75	Slimline	WA 296 A	—	22

G

Submittal Package—Lighting Fixtures
Figure 13-3 (continued)

MOTEL/HOTEL LIGHTS
Incandescent or fluorescent

All aluminum construction — Asymmetric lenses for controlled illumination. — Two compartment lighting for separately controlled indirect up-lighting and direct down-lighting — Voltage must be specified on fluorescent units.

TYPE H

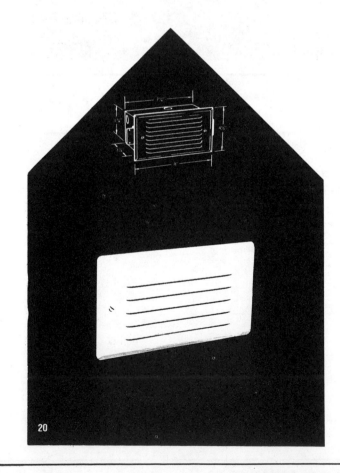

Cat. No.	Wattage	Dimensions A	B	Ballast
Incandescent				
B101	2-75W	28"	25 ½"	
B102	4-60W	41"	38 ½"	
B103	4-75W	53"	50 ½"	
Fluorescent				
B105	1-20WT12	28"	25 ½"	1
B106	2-20WT12	28"	25 ½"	1
B107*	2-20WT12	28"	25 ½"	2
B112	1-30WT12	41"	38 ½"	1HPF
B122	2-30WT12	41"	38 ½"	1HPF
B132* *H*	2-30WT12	41"	38 ½"	2HPF
B113	1-40WT12	53"	50 ½"	1HPF
B123	2-40WT12	53"	50 ½"	1HPF
B133*	2-40WT12	53"	50 ½"	2HPF

*Pull chain for downlighting standard
Finish must be specified:
W walnut vinyl   **V** linen vinyl   **P** white paint
Optional accessories:
/20 grounded convenience outlet
/23 pull chain switch (when not standard)

A - overall length
B - length of housing

25 watt night light — Recessed housing with aluminum louvered face panel — low gloss white painted finish — Optional pin switch available, specify **/21** after catalog number.

Cat. No.	Description
NL-IP	Nite Lite, complete unit

Submittal Package—Lighting Fixtures
Figure 13-3 (continued)

M1 M2 M3

SQUARES
100 - 150 - 200 WATT

Type J

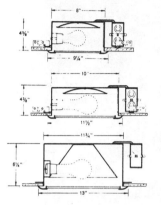

	Standard trim, only	Complete	Multi-step trim, only	Complete
Flat Albalite	10	M1-10	F10	M1-F10
Drop Opal	11	M1-11 ✓ J	F11	M1-F11
Flat Fresnel	16	M1-16	F16	M1-F16
Opalene*	P11	M1-P11		
Flat Albalite	20	M2-20	F20	M2-F20
Drop Opal	21	M2-21	F21	M2-F21
Flat Fresnel	26	M2-26	F26	M2-F26
Flat Albalite	280	M3-280	F280	M3-F280
Drop Opal	281	M3-281	F281	M3-F281
Flat Fresnel	286	M3-286	F286	M3-F286

*opalene is an injection molded improved polyproprylene

I.B.E.W Union made   U.L. Listed for damp locations

Finish must be specified: **A** aluminum
P white paint, **C** chrome, **R** brass

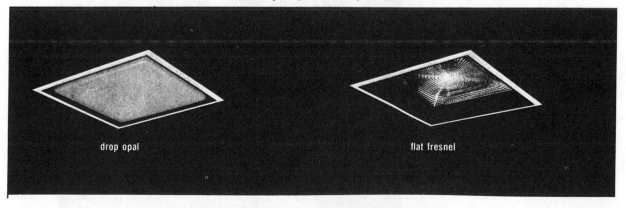

drop opal

flat fresnel

Submittal Package—Lighting Fixtures
Figure 13-3 (continued)

Johns-Manville
JM

Holophane®
Lobay® II
luminaires.

TYPE K

Square distribution,
wide spacing.

Low and medium
mounting heights.

HPS, metal halide
or mercury.

Holophane
HL-506
11
78
16p
LIGHTING
industrial

Provided by the Holophane Division of Johns-Manville Sales Corporation

Socket Series
Industrial lighting fixtures for 100W thru 500W incandescent lamps.

TYPE L

Ordering Information and Dimensions
Fixture Catalog numbers include reflector, hood, and standard socket◊†

Lamp Watts	Dia.(A) Inches	Hght.(B) Inches	Catalog Numbers Reflector Only (Less Socket)	With Pendant Socket◊†
Standard Dome Fixtures*				
100	12	7-9/16	G-50012	G-50212
150	14	8-1/4	G-50014	G-50214
200	16	9-1/2	G-50016	G-50216
300-500	18	11-5/8	G-50018	G-50218
Symmetrical Angle Fixtures				
100	8	8-7/8	G-90028	G-90228
150	10	11-5/8	G-90010	G-90210
200	12	12-5/8	G-90012	G-90212
300-500	14	15-3/4	G-90014	G-90214

Yard Light
For 150W medium base lamps. Includes shallow dome porcelain enameled reflector, white inside and out; cast aluminum hood with pre-wired socket; 1/2" x 15" plated conduit stem, insulated flange, 7" exposed leads.

G-17244

Heavy-Duty Deep Wire Guards

14	5	G-1314
16	5-1/4	G-1316
18	5-1/4	G-1318

Pendant Sockets, Keyless, 600V

600W Medium Base	1/2"◊	G-53
1500W Mogul Base	1/2"◊	G-63

Pendant Shock-Absorbing Sockets, Keyless, 600V

600W Medium Base	1/2"◊	G-52
1500W Mogul Base	1/2"◊	G-62

DISCOUNT SCHEDULE GL-R

Square Outlet Box Cover with 1/2" KO.
For installing fixture to 4" square outlet box.
Requires CN-50 Conduit Nipple.

8474

Octagonal Outlet Box Cover with 1/2" KO.
For installing fixture to 4" octagonal outlet box.
Requires CN-50 Conduit Nipple.

8413

DISCOUNT SCHEDULE OB

Conduit Nipple
For use with outlet box covers above.

CN-50

DISCOUNT SCHEDULE CF-2

*Standard Dome fixture only is also available with ventilated reflector. To order, add suffix **—V** to catalog number. ◊Sockets tapped 1/2" standard. Add suffix **—3/4** if 3/4" tapped hub is desired. †Catalog numbers are for fixture with standard socket. To order reflector with shock-absorbing socket, add suffix **—SA** to catalog number.

Discount Schedule (See Above)
Refer to Pricing Index for price,
weight, and standard package

PAGE 10, Effective 1980

Appleton ELECTRIC COMPANY

1701 W. Wellington Ave.
Chicago, Illinois 60657

Copyright 1980 Printed in U.S.A.

Submittal Package—Lighting Fixtures
Figure 13-3 (continued)

MRF 400
Floodlighting

Ordering Information

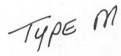

TYPE M

✓M

Lamp Type	Wattage	Ballast Type	Voltage①	Power Factor	NEMA Beam Spread	Weight In Lbs.	Photocontrol Receptacle	Catalog Number②
HPS	150③	CW	120	High	7 x 6	60	Yes	FR4-GSJG-A76A
HPS	250	CW	120/208/240/277	High	7 x 6	65	Yes	FR4-GSNG-W76A
HPS	400	CW	120/208/240/277	High	7 x 6	65	Yes	FR4-GSPG-W76A
Metal Halide	250	CWA	120/208/240/277	High	7 x 6	53	Yes	FR4-GJNA-W76A
Metal Halide	250	CWA	120	High	7 x 6	51	Yes	FR4-GJNA-A76A
Metal Halide	400	CWA	120/208/240/277	High	7 x 6	53	Yes	FR4-GJPA-W76A
Metal Halide	400	CWA	120/240	High	7 x 6	51	Yes	FR4-GJPA-F76A
Mercury	250	CWA	120/208/240/277	High	7 x 6	51	Yes	FR4-GDNA-W76A
Mercury	400	CWA	120/208/240/277	High	7 x 6	51	Yes	FR4-GDPA-W76A
Mercury	400	CWA	120/240	High	7 x 6	48	Yes	FR4-GDPA-F76A

Modifications: To order for the above fixture types, add 6 digit suffix to catalog number.

Suffix Adder④	Description
ABAAEA	Four foot length of 3 conductor #14 SO cord attached at factory
ABBAAA	Single 600 volt KTK fuse and holder—In Line
ABBAEA	Both four foot 3 conductor #14 SO cord and single 600 volt KTK fuse and holder

① Other voltages available on request.
② Luminaires cataloged below do not include lamp or photocontrol.
③ 100 volt lamp.
④ Minimum order quantity required. Contact factory.

Photometric Data: MRF 400 Cat. No. FR4-GDPA-W76A

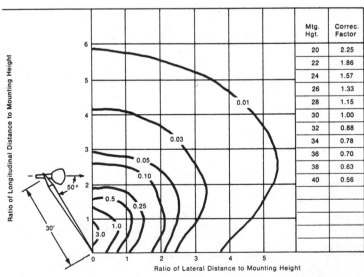

Mtg. Hgt.	Correc. Factor
20	2.25
22	1.86
24	1.57
26	1.33
28	1.15
30	1.00
32	0.88
34	0.78
36	0.70
38	0.63
40	0.56

Submittal Package—Lighting Fixtures
Figure 13-3 (continued)

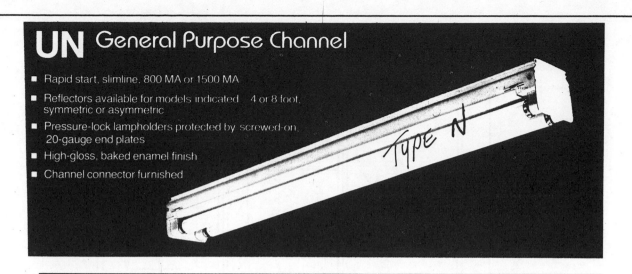

UN General Purpose Channel

- Rapid start, slimline, 800 MA or 1500 MA

- Reflectors available for models indicated—4 or 8 foot, symmetric or asymmetric

- Pressure-lock lampholders protected by screwed-on, 20-gauge end plates

- High-gloss, baked enamel finish

- Channel connector furnished

Type N

NARROW BODY 5″
4, 8 or 16 Foot Length,
1, 2 or 4 Lamps

  4-9/16″ 5″

WIDE BODY 9″
4, 8 or 16 Foot Length,
2, 3, 4 or 6 Lamps

    4-9/16″ 9″

ORDERING INFORMATION

Lamps (Not Included) Type	No.-Watts	Nominal Length (Feet)	Catalog Number	Dimensions (Inches) Length	Width	Height	Channel Dia. No. (Page 110)
Rapid Start	1-40	4	UN 140	48	5	4-9/16	38
Rapid Start	1-40	4	UN 140	48	5	4-9/16	38
Rapid Start	2-40	4	2UN 240	48	9	4-9/16	38
Rapid Start	3-40	4	UN 340	48	9	4-9/16	38
Rapid Start	4-40	4	UN 440	48	9	4-9/16	38
Rapid Start	2-40	8	8TUN 140	96	5	4-9/16	38
Rapid Start	4-40	8	8TUN 240	96	5	4-9/16	38
Slimline	1-38	4	UN 148	48	5	4-9/16	32
Slimline	2-38	4	UN 248	48	5	4-9/16	32
Slimline	2-38	4	2UN 248	48	9	4-9/16	32
Slimline	3-38	4	UN 348	48	9	4-9/10	32
Slimline	4-38	4	UN 448	48	9	4-9/16	32
Slimline	1-75	8	UN 196	96	5	4-9/16	39
Slimline	2-75	8	UN 296	96	5	4-9/16	39
Slimline	2-75	8	2UN 296	96	9	4-9/16	39
Slimline	3-75	8	UN 396	96	9	4-9/16	39
Slimline	4-75	8	UN 496	96	9	4-9/16	39
Slimline	2-75	16	8TUN 396	192	5	4-9/16	39
Slimline	6-75	16	8TUN 396	192	9	4-9/16	39
800 MA	1-60	4	UN 148 HO	48	5	4-9/16	42
800 MA	2-60	4	UN 248 HO	48	5	4-9/16	42
800 MA	2-60	4	2UN 248 HO	48	9	4-9/16	42
800 MA	1-110	8	UN 196 HO	96	5	4-9/16	39
800 MA	2-110	8	UN 296 HO	96	5	4-9/16	39
800 MA	2-110	8	2UN 296 HO	96	9	4-9/16	39
800 MA	2-110	16	8TUN 196 HO	192	5	4-9/16	39
1500 MA	1-110	4	UN 148 PG	48	5	4-9/16	42
1500 MA	2-110	4	UN 248 PG	48	5	4-9/16	42
1500 MA	1-215	8	UN 196 PG	96	5	4-9/16	39
1500 MA	2-215	8	UN 296 PG	96	5	4-9/16	39
1500 MA	2-215	16	8TUN 196 PG	192	5	4-9/16	39

Optional Reflectors   Order Separately

Catalog Number Asymmetric	Symmetric	Description
AS 48	EJR48	Reflector for 4-foot models
AS 96	EJR 96	Reflector for 8-foot models

Submittal Package—Lighting Fixtures
Figure 13-3 (continued)

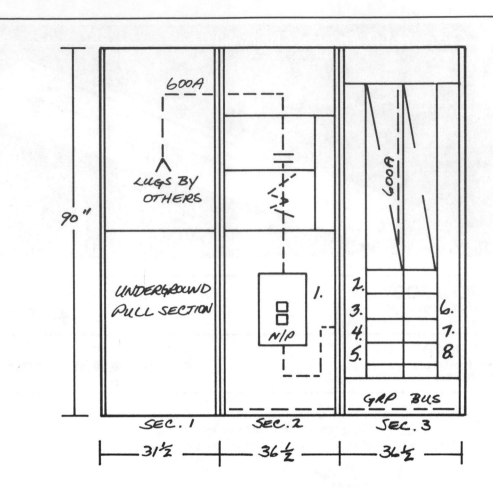

Submittal Package—Shop Drawing—Switchboard
Figure 13-4

ITEM	QTY	CAT. NO.	BILL of Material	
1.	1	NN630600	600A 3P NN BREAKER	SERVICE DISCONNECT
2.	1	NTL630150	150A 3P NTL-6 BREAKER	PANEL A
3.	1	NTL630125	125A 3P NTL-6 BREAKER	PANEL B
4.	1	HEJ230090	90A 3P HEJ BREAKER	PANEL C
5.	1	HEJ230070	70A 3P HEJ BREAKER	PANEL D
6.	1	HEJ230150	150A 3P HEJ BREAKER	SPARE
7.	1	HEJ230100	100A 3P HEJ BREAKER	SPARE
8.	1	HEJ230100	100A 3P HEJ BREAKER	SPARE

PROJECT: ADVERTISING COMPANY
DESIGNATION: SWITCHBOARD "MSB"

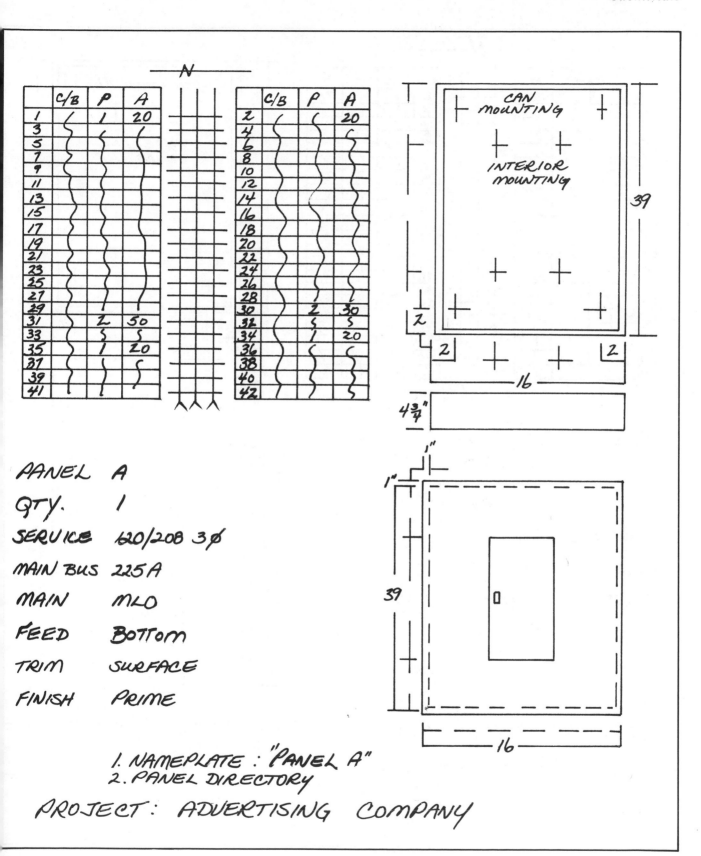

	C/B	P	A
1	(	1	20
3			
5			
7			
9			
11			
13			
15			
17			
19			
21			
23			
25			
27			
29			
31		2	50
33			
35		1	20
37			
39			
41			

	C/B	P	A
2	(	(	20
4			
6			
8			
10			
12			
14			
16			
18			
20			
22			
24			
26			
28			
30		2	30
32			
34		1	20
36			
38			
40			
42			

PANEL   A

QTY.   1

SERVICE   120/208 3φ

MAIN BUS 225A

MAIN   MLO

FEED   BOTTOM

TRIM   SURFACE

FINISH   PRIME

1. NAMEPLATE : "PANEL A"
2. PANEL DIRECTORY

PROJECT : ADVERTISING COMPANY

Submittal Package—Panelboard
Figure 13-5

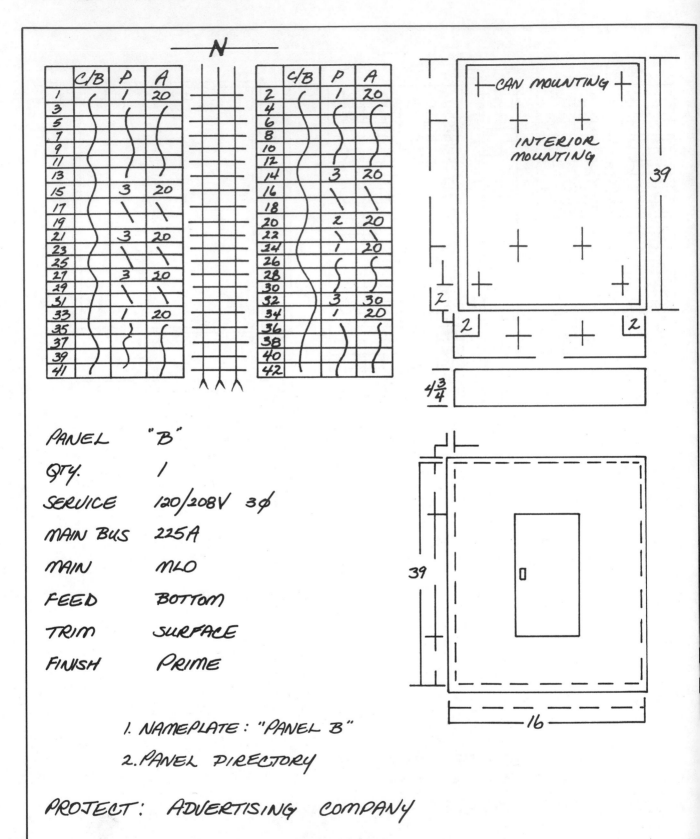

	C/B	P	A
1		1	20
3			
5			
7			
9			
11			
13			
15		3	20
17			
19			
21		3	20
23			
25			
27		3	20
29			
31			
33		1	20
35			
37			
39			
41			

	C/B	P	A
2		1	20
4			
6			
8			
10			
12			
14		3	20
16			
18			
20		2	20
22			
24		1	20
26			
28			
30			
32		3	30
34		1	20
36			
38			
40			
42			

N

CAN MOUNTING

INTERIOR MOUNTING

39

2   2   2

4 3/4

39

16

PANEL   "B"

QTY.   1

SERVICE   120/208V 3φ

MAIN BUS   225A

MAIN   MLO

FEED   BOTTOM

TRIM   SURFACE

FINISH   PRIME

1. NAMEPLATE: "PANEL B"

2. PANEL DIRECTORY

PROJECT: ADVERTISING COMPANY

Submittal Package—Panelboard
Figure 13-5 (continued)

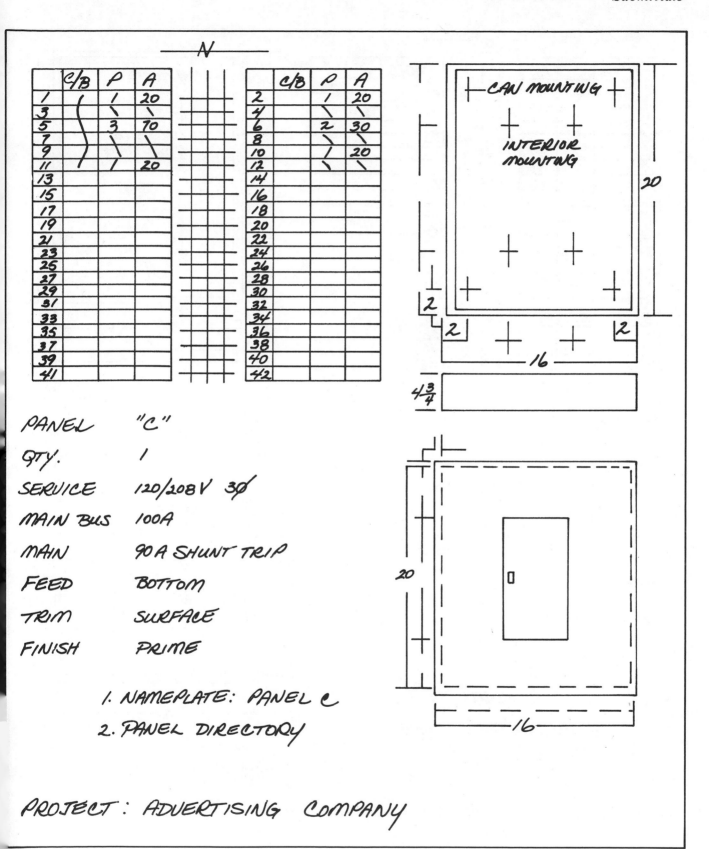

	C/B	P	A				C/B	P	A
1		1	20			2		1	20
3						4			
5		3	70			6		2	30
7						8			
9						10		1	20
11		1	20			12			
13						14			
15						16			
17						18			
19						20			
21						22			
23						24			
25						26			
27						28			
29						30			
31						32			
33						34			
35						36			
37						38			
39						40			
41						42			

PANEL   "C"

QTY.   1

SERVICE   120/208V 3∅

MAIN BUS   100A

MAIN   90A SHUNT TRIP

FEED   BOTTOM

TRIM   SURFACE

FINISH   PRIME

1. NAMEPLATE: PANEL C

2. PANEL DIRECTORY

PROJECT: ADVERTISING COMPANY

Submittal Package—Panelboard
Figure 13-5 (continued)

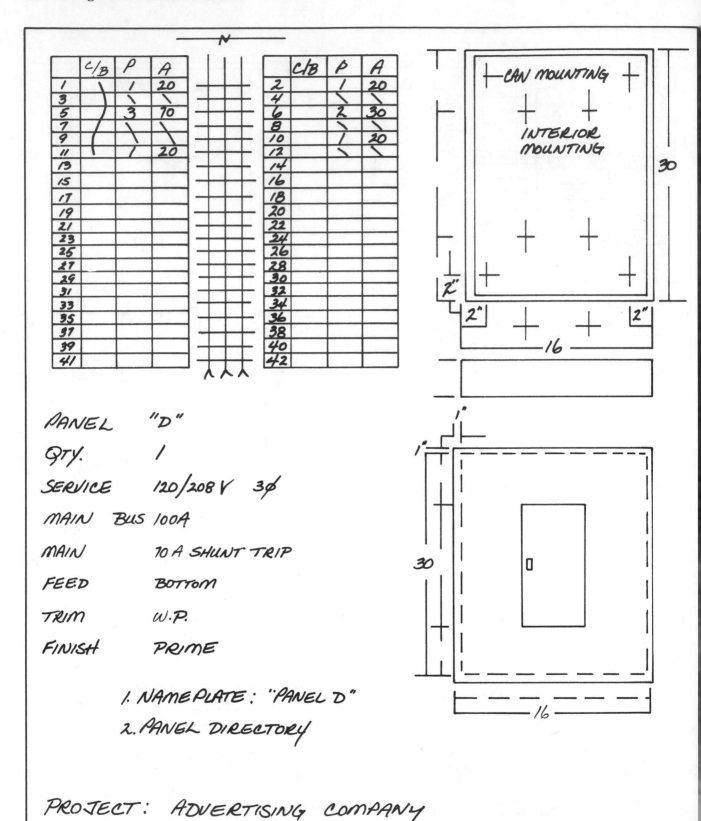

PANEL   "D"

QTY.   1

SERVICE   120/208 V 3∅

MAIN BUS 100A

MAIN   70 A SHUNT TRIP

FEED   BOTTOM

TRIM   W.P.

FINISH   PRIME

1. NAMEPLATE: "PANEL D"
2. PANEL DIRECTORY

PROJECT: ADVERTISING COMPANY

Submittal Package—Panelboard
Figure 13-5 (continued)

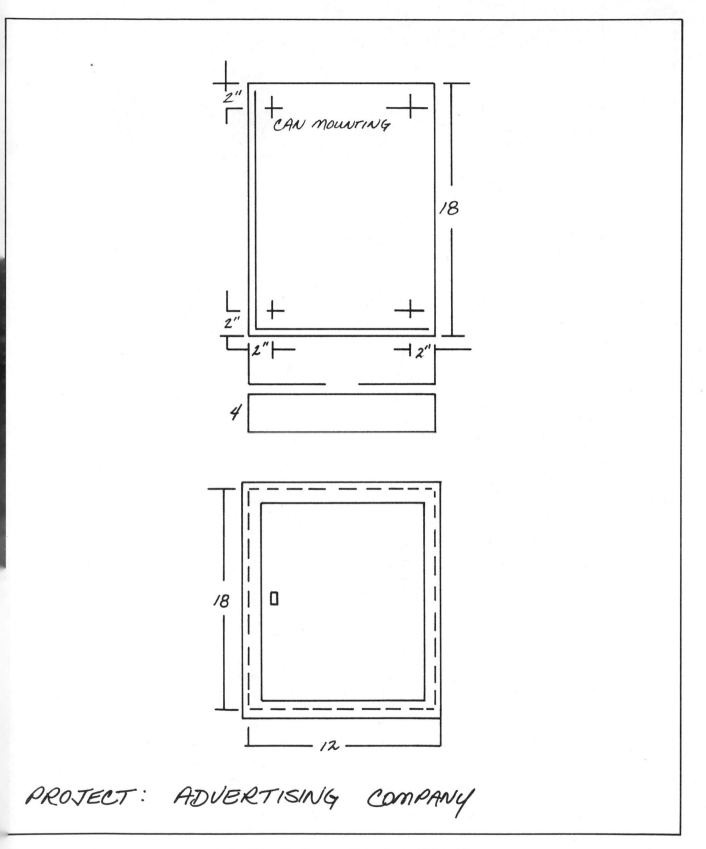

Submittal Package—Telephone Cabinet
Figure 13-6

SAFETY SWITCHES — HEAVY DUTY
VISIBLE BLADES

NEMA 1    NEMA 3R 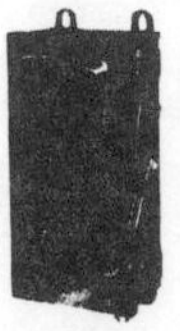   NEMA 4 and 5 Stainless Steel   NEMA 12

Visible Blade Heavy Duty Safety Switches are designed for application where maximum performance and continuity of service are required. All Heavy Duty Safety Switches feature Quick-make, Quick-break operating mechanism, full cover interlock and indicator handle. They meet Federal Specification W-S-865c for Heavy Duty Switches, are UL listed (File E2875), and meet NEMA KSI-1975 for Type HD. Suitable for use as service equipment.
Finish—Gray baked enamel over rust inhibiting primer.

240 VOLT — SINGLE THROW FUSIBLE

Systems	Amps.	NEMA 1 Indoor Cat. No.	Price	NEMA 3R Rainproof (Bolt-on Hubs Page 41) Cat. No.	Price	NEMA 4 and 5 Dusttight, Watertight Stainless Steel (Hubs Page 43) Cat. No.	Price	JIC-Mill & Foundry Type NEMA 12K With Knockouts Cat. No.	NEMA 12 Without Knockouts Cat. No.	Price	HP 240 V. ac Std. 1φ	3φ	Max. 1φ	3φ	250 V. dc	Amps.
2 POLE, 240 VOLTS AC — 250 VOLTS DC																
	30	H221	$ 61.	Use 2 Pole— 3 Wire For 2 Pole Application		H221DS	$ 466.	H221A	H221AWK	$ 114.	1½	..	3	..	5	30
	30	✱H221-2	105.					✱H221-2A	✱H221-2AWK	141.	1½	..	3	..	5	30
	60	H222	118.			H222DS	501.	H222A	H222AWK	156.	3	..	10	..	10	60
	100	H223	183.			H223DS	1230.	H223A	H223AWK	225.	7½	..	15	..	20	100
	200	H224	328.	H224RB	$ 391.	H224DS	1685.	H224A	H224AWK	379.	15	..	..	..	40	200
	400	H225	679.	H225R	968.	H225DS	3427.	H225A	H225AWK	865.	..	..	..	..	50	400
	600	H226	1351.	H226R	1814.	H226DS	4919.	......	H226AWK	1493.	..	..	..	..	..	600
	800	△H227	2094.	△H227R	2844.	....	....		‡	....	..	..	..	..	..	....
	1200	△H228	2892.	△H228R	3835.	....	....		‡	....	..	..	..	..	..	....
3 WIRE S/N (2 BLADES 2 FUSES) 240 VOLTS AC — 125/250 VOLTS DC																
	30	H221N	$ 61.	H221NRB	$ 114.	H221NDS	$ 483.	H221NA	H221NAWK	$ 126.	1½	3	3	7½	5	30
	60	H222N	118.	H222NRB	214.	H222NDS	585.	H222NA	H222NAWK	167.	3	7½	10	15	10	60
	100	H223N	183.	H223NRB	276.	H223NDS	1269.	H223NA	H223NAWK	268.	7½	15	15	30	20	100
	200	H224N	328.	H224NRB	398.	H224NDS	1739.	H224NA	H224NAWK	419.	15	25	..	60	40	200
	400	H225N	771.	H225NR	968.	H225NDS	3499.	......	H225NAWK	1021.	..	50	..	125	50	400
	600	H226N	1450.	H226NR	1814.	H226NDS	4991.	......	H226NAWK	1583.	..	75	..	200	..	600
	800	△H227N	2492.	△H227NR	3024.	....	....		‡	....	..	..	..	..	..	....
	1200	△H228N	3076.	△H228NR	4125.	....	....		‡	....	..	..	..	..	..	....
3 POLE, 240 VOLTS AC																
	30	H321	$ 80.	H321RB	$ 141.	H321DS	$ 491.	H321A	H321AWK	$ 141.	..	3	..	7½		30
	30	✱H321-2	126.					✱H321-2A	✱H321-2AWK	167.	..	3	..	7½		30
	60	H322	135.	H322RB	227.	H322DS	608.	H322A	H322AWK	199.	..	7½	..	15		60
	100	H323	214.	H323RB	326.	H323DS	1293.	H323A	H323AWK	308.	..	15	..	30		100
	200	H324	370.	H324RB	444.	H324DS	1815.	H324A	H324AWK	463.	..	25	..	60		200
	400	H325	852.	H325R	990.	H325DS	3540.	H325A	H325AWK	1009.	..	50	..	125		400
	600	H326	1537.	H326R	2064.	H326DS	5067.	......	H326AWK	1679.	..	75	..	200		600
	800	H327	2836.	H327R	3675.	....	....		‡	....	..	..	..	..		....
	1200	H328	3599.	H328R	4794.	....	....		‡	....	..	..	..	..		....
4 WIRE S/N (3 BLADES 3 FUSES) 240 VOLTS AC																
	30	H321N	$ 80.	H321NRB	$ 141.	H321NDS	$ 514.	H321NA	H321NAWK	$ 159.	..	3	..	7½	..	30
	60	H322N	135.	H322NRB	227.	H322NDS	628.	H322NA	H322NAWK	214.	..	7½	..	15	..	60
	100	H323N	214.	H323NRB	326.	H323NDS	1332.	H323NA	H323NAWK	347.	..	15	..	30	..	100
	200	H324N	370.	H324NRB	444.	H324NDS	1872.	H324NA	H324NAWK	499.	..	25	..	60	..	200
	400	H325N	944.	H325NR	1078.	H325NDS	3540.	H325NA	H325NAWK	1099.	..	50	..	125	..	400
	600	H326N	1625.	H326NR	2149.	H326NDS	5067.	......	H326NAWK	1769.	..	75	..	200	..	600
	800	H327N	3017.	H327NR	3852.	....	....		‡	....	..	..	..	..	..	....
	1200	H328N	3792.	H328NR	4975.	....	....		‡	....	..	..	..	..	..	....
4 POLE, 240 VOLTS AC △											2φ		2φ			
	30	✱H421-2	$ 156.	....	....	....	....		✱H421-2AWK	$199.	..	3	..	10		30
	60	H422	210.	....	....	....	....		H422AWK	242.	..	7½	..	20		60
	100	H423	328.	....	....	....	....		H423AWK	400.	..	15	..	30		100
	200	H424	594.	....	....	....	....		H424AWK	698.	..	30	..	60		200
	400	H425	1134.	....	....	....	....		H425AWK	1339.	..	..	..	60		400
	600	H426	2028.	....	....	....	....				..	..	..	..		600

✱60 ampere switch with 30 ampere fuse spacing and clips.
‡For application above 600 amperes, refer to BOLT-LOC switches on page 51.
△AC only.
●The starting current of motors of more than standard horsepower rating may require the use of fuses with appropriate time delay characteristics.
▲Not suitable for use as service equipment.

Electrical Interlocks—Pages 44, 46 & 47
Dimensions: NEMA 1 & 3R—Page 46
NEMA 4 & 5, 12—Page 47

BOLT-LOC is a Registered Trademark of Square D Company.

CLASS R FUSE KITS—240 Volt (One kit per 3 pole switch)

Amps.	Series	Catalog Number	Price Each
30	D2 & D4	HRK-30◑	$ 6.10
30	D1	HTK-30	6.10
60	D1	HTK-60	6.10
60	D2 & D4	HRK-60	6.10
100	D2, D3 & D4	HRK-1020	11.20
200	D2 & D4	HRK-1020	11.20
400	D2	HRK-4060	26.00
600	D2	HRK-4060	26.00

Kits reject all except class R fuses. Class R fuse will install in standard switch **without** the use of kit. Kit is to reject class H, K, etc. fuses.
◑H221-2, H321-2, H421-2, H221-2A,AWK, H321-2A,AWK and H421-2AWK use QMB-30R Class R fuse kit.
Price **$11.20 G1-1 DISCOUNT.**

Reprinted with permission of Square D Company

Submittal Package—Catalog Data
Figure 13-7

Non-Fusible Combination Starter in NEMA 1 Enclosure

AC COMBINATION STARTERS
NON-FUSIBLE DISCONNECT SWITCH TYPE

CLASS 8538

The disconnect switch type combination starter design utilizes a flange operated visible blade switch. Size 0-2, non-fusible combination starters can be converted to a fusible type. All NEMA Type 1, 4 and 12 enclosures have the same size enclosure for fusible and non-fusible combination starters. Therefore, space is available for the conversion to a fusible type.

LINE VOLTAGE TYPE, NON-REVERSING WITH THREE MELTING ALLOY OVERLOAD RELAYS
3 POLE — 600 VOLTS MAX. — 50-60 HERTZ

Motor Voltage (Starter Voltage)	Max. HP Poly-phase	NEMA Size	General Purpose Enclosure NEMA Type 1		Watertight and Dusttight Enclosure Stainless Steel (Sizes 0-5) NEMA Type 4		Watertight, Dusttight and Corrosion Resistant Enclosure NEMA Type 4X▲		Dusttight and Driptight, Industrial Enclosure NEMA Type 12◐		
									With External Reset	Without External Reset	
			Type	Price*	Type	Price*	Type	Price*	Type	Type	Price*
200 (208)	3	0	SBG-11	$ 364.	SBW-11	$ 748.	SBW-21	$ 860.	SBA-21	SBA-11	$ 460.
	7½	1	SCG-11	384.	SCW-11	768.	SCW-21	884.	SCA-21	SCA-11	480.
	10	2	SDG-11	608.	SDW-11	1200.	SDW-21	1320.	SDA-21	SDA-11	744.
	25	3	SEG-11	1012.	SEW-11	2044.	SEW-21	2248.	SEA-21	SEA-11	1188.
	40	4	SFG-11	1952.	SFW-11	3272.	....	....	SFA-21	SFA-11	2440.
	75	5	SGG-11	4382.	SGW-11	7750.	....	....	SGA-21	SGA-11	5566.
	150	6	SHG-11	11562.	SHW-11	14762.	....	....	SHA-21	SHA-11	12942.
230 (240)	3	0	SBG-11	364.	SBW-11	748.	SBW-21	860.	SBA-21	SBA-11	460.
	7½	1	SCG-11	384.	SCW-11	768.	SCW-21	884.	SCA-21	SCA-11	480.
	15	2	SDG-11	608.	SDW-11	1200.	SDW-21	1320.	SDA-21	SDA-11	744.
	30	3	SEG-11	1012.	SEW-11	2044.	SEW-21	2248.	SEA-21	SEA-11	1188.
	50	4	SFG-11	1952.	SFW-11	3272.	....	....	SFA-21	SFA-11	2440.
	100	5	SGG-11	4382.	SGW-11	7750.	....	....	SGA-21	SGA-11	5566.
	200	6	SHG-11	11562.	SHW-11	14762.	....	....	SHA-21	SHA-11	12942.
460-575 (480-600)	5	0	SBG-11	364.	SBW-11	748.	SBW-21	860.	SBA-21	SBA-11	460.
	10	1	SCG-11	384.	SCW-11	768.	SCW-21	884.	SCA-21	SCA-11	480.
	25	2	SDG-11	608.	SDW-11	1200.	SDW-21	1320.	SDA-21	SDA-11	744.
	50	3	SEG-11	1012.	SEW-11	2044.	SEW-21	2248.	SEA-21	SEA-11	1188.
	100	4	SFG-11	1952.	SFW-11	3272.	....	....	SFA-21	SFA-11	2440.
	200	5	SGG-11	4382.	SGW-11	7750.	....	....	SGA-21	SGA-11	5566.
	400	6	SHG-11	11562.	SHW-11	14762.	....	....	SHA-21	SHA-11	12942.

* Prices do not include thermal units. For selection, see page 436.
◐ NEMA Type 12 enclosures may be field modified for outdoor applications. For details refer to page 418.
▲ NEMA Type 4X hubs are included with each starter at no additional cost.

UL LISTED FOR 100,000A ‡

AC COMBINATION STARTERS
FUSIBLE DISCONNECT SWITCH TYPE WITH CLASS R FUSE CLIPS

LINE VOLTAGE TYPE, NON-REVERSING WITH THREE MELTING ALLOY OVERLOAD RELAYS
3 POLE — 600 VOLTS MAX. — 50-60 HERTZ

Motor Voltage (Starter Voltage)	Max. HP Poly-phase	NEMA Size	Fuse Clip Size Amps.	General Purpose Enclosure NEMA Type 1		Watertight and Dusttight Enclosure Stainless Steel (Sizes 0-5) NEMA Type 4		Watertight, Dusttight and Corrosion Resistant Enclosure NEMA Type 4X▲		Dusttight and Driptight, Industrial Enclosure NEMA Type 12◐		
										With External Reset	Without External Reset	
				Type	Price*	Type	Price*	Type	Price*	Type	Type	Price*
200 (208)	3	0	30	SBG-32	$ 382.	SBW-32	$ 766.	SBW-42	$ 880.	SBA-42	SBA-32	$ 478.
	5	1	30	SCG-32	402.	SCW-32	786.	SCW-42	904.	SCA-42	SCA-32	498.
	7½	1	60	SCG-33	410.	SCW-33	794.	SCW-43	912.	SCA-43	SCA-33	506.
	10	2	60	SDG-32	630.	SDW-32	1222.	SDW-42	1344.	SDA-42	SDA-32	766.
	20	3	100	SEG-35	1064.	SEW-35	2096.	SEW-45	2306.	SEA-45	SEA-35	1240.
	25	3	200	SEG-32	1152.	SEW-32	2184.	....	....	SEA-42	SEA-32	1328.
	40	4	200	SFG-35	2032.	SFW-35	3352.	....	....	SFA-45	SFA-35	2520.
	75	5	400	SGG-35	4554.	SGW-35	7922.	....	....	SGA-45	SGA-35	5738.
	150	6	600	SHG-33	12014.	SHW-33	15214.	....	....	SHA-43	SHA-33	13394.
230 (240)	3	0	30	SBG-32	382.	SBW-32	766.	SBW-42	880.	SBA-42	SBA-32	478.
	5	1	30	SCG-32	402.	SCW-32	786.	SCW-42	904.	SCA-42	SCA-32	498.
	7½	1	60	SCG-33	410.	SCW-33	794.	SCW-43	912.	SCA-43	SCA-33	506.
	15	2	60	SDG-32	630.	SDW-32	1222.	SDW-42	1344.	SDA-42	SDA-32	766.
	25	3	100	SEG-35	1064.	SEW-35	2096.	SEW-45	2306.	SEA-45	SEA-35	1240.
	30	3	200	SEG-32	1152.	SEW-32	2184.	....	....	SEA-42	SEA-32	1328.
	50	4	200	SFG-35	2032.	SFW-35	3352.	....	....	SFA-45	SFA-35	2520.
	100	5	400	SGG-35	4554.	SGW-35	7922.	....	....	SGA-45	SGA-35	5738.
	200	6	600	SHG-33	12014.	SHW-33	15214.	....	....	SHA-43	SHA-33	13394.
460-575 (480-600)	5	0	30	SBG-33	390.	SBW-33	774.	SBW-43	890.	SBA-43	SBA-33	486.
	10	1	30	SCG-34	410.	SCW-34	794.	SCW-44	912.	SCA-44	SCA-34	506.
	15	2	30	SDG-36	634.	SDW-34	1226.	SDW-46	1348.	SDA-46	SDA-36	770.
	25	2	60	SDG-33	642.	SDW-34	1234.	SDW-44	1358.	SDA-44	SDA-34	778.
	50	3	100	SEG-33	1084.	SEW-33	2116.	SEW-43	2328.	SEA-43	SEA-33	1260.
	100	4	200	SFG-33	2048.	SFW-33	3368.	....	....	SFA-43	SFA-33	2536.
	200	5	400	SGG-33	4554.	SGW-33	7922.	....	....	SGA-43	SGA-33	5738.
	400	6	600	SHG-32	12014.	SHW-32	15214.	....	....	SHA-42	SHA-32	13394.

* **Prices do not include thermal units. For selection, see page 436.**
◐ NEMA Type 12 enclosures may be field modified for outdoor applications. For details refer to page 418.
▲ NEMA Type 4X hubs are included with each starter at no additional cost.
‡ Available amperes RMS symmetrical. NEMA Type 4X devices are not UL Listed.

Ordering Information — Refer to Page 226.
Field modification kits — Refer to Pages 228-230 and Class 9999 section.

Submittal Package—Catalog Data
Figure 13-7 (continued)

Chapter 14

Change Orders and Backcharges

Change Orders

A change order is a written notice to the contractor authorizing specific changes in the work under contract. It can only be issued by the owner or his representative during the term of the contract. The order should describe in detail the work to be changed. It should give the amount to be added to or deducted from the contract price.

Some change orders have no effect on the contract price. These change orders merely establish for the record that a specific change was made.

Changes made before the bid is accepted can be added to the bid as an *addendum*. The architect distributes the addenda to the bidding contractors, who then adjust the contract documents accordingly.

But changes made *after* the bid has been accepted and the contract has been executed must be handled by change orders. Most contracts have provisions for change orders. A typical contract provision is shown in Figure 14-1.

Carefully check each contract change order provision before preparing the change order cost breakdown. Provisions vary from contract to contract. Some are standard forms developed by architects' associations. Others were developed by contractors' associations. And some were developed by individual contractors. All have been reviewed by attorneys, and many have been tested in court.

Why Changes are Made

Usually, construction documents are drafted and assembled long before the job is let out for bid. During this time it may become necessary to change or revise certain parts of the documents. The changes may or may not seriously affect the plans.

There are at least four main reasons for change orders. They are:
1. Requirements by the authorities.
(a) Building code compliance.
(b) Fire code compliance.
(c) Zoning compliance.
(d) Utility companies compliance.
2. Financial impact.
(a) Insufficient funding.
(b) Additional funding.
(c) Building budget reserves.
3. Owner's benefit.
(a) Tenant needs.

Changes In The Work

The Subcontractor hereby agrees that the Contractor or his Superintendent shall have the right to order, in writing, additions or deletions to the Contract as may be required throughout the term of the Contract. The Subcontractor shall make all changes in accordance with the Contract scope of work. No changes in the work shall be made without the written authorization from the Contractor. No payment shall be made for changes, unless prior written approval from the Contractor.

The Contractor shall select one of the following methods in determining the charge or credit for each written Change Order:

1. The written Change Order or Field Change Order shall stipulate an agreed LUMP SUM PRICE to be added or to be deducted from the Contract.

2. The Change Order shall be estimated by the Subcontractor and a complete cost breakdown listing all material, equipment, delivery and labor cost or credit shall be submitted to the Contractor for approval. If the change is in addition to the Contract Price, the Subcontractor may charge overhead at 15% and profit at 10%. If the change amounts to a credit, no overhead or profit will be included.

3. The Contractor may order the Subcontractor to proceed with the change and to promptly submit a cost breakdown listing all material, equipment, delivery and labor cost or credit. If the change is in addition to the Contract Price, the Subcontractor may charge overhead at 15% and profit at 10%. If the change amounts to a credit, no overhead or profit will be included.

4. Any change in the work done without authorization may be subject to corrective measure without cost to the Contractor. The Contractor may order the Subcontractor to remove unauthorized work at no cost to the Contract. Any unauthorized work may be accepted by the Contractor at no change in Contract price if found to be in the interest of the Owner.

5. The Subcontractor shall include in the written cost breakdown any additions to the Contract completion time that might be required. The amount of time shall be listed in calendar days.

Typical Contract Provision
Figure 14-1

(b) Contract unit cost changes.

(c) New technology.

(d) New construction methods or materials.

4. Design deficiencies.

(a) Design errors or omissions.

(b) Poor coordination within the design team.

(c) Erroneous site data.

Of the above, design deficiencies are the most common cause of change orders. Basically there are two reasons for this: design errors and poor communication between the owner and the architect.

Design errors can happen when details and specifications are borrowed from similar projects. The owner might take this shortcut to reduce the cost of the design work. But studies show that this practice causes up to 50 percent of all change orders. Nationally, this can amount to millions of dollars in changes each year. A set of specifications specially designed for a project can reduce the number of change orders.

The other major cause of change orders is misinterpretation of the owner's requirements by the architect. The architect may not fully understand what the owner wants. Or the owner may not understand what the architect has designed. This isn't your problem until someone discovers that a change has to be made. Construction is too permanent and too expensive to let a serious flaw go uncorrected.

Most contractors are happy to handle changes. The work is usually done at your price and isn't subject to competitive bid. But the trick is to recover all your costs on changes and make a reasonable profit. Most important, don't handle a change without first receiving a written promise of payment for the change. It's too easy to forget what change was made and how much it was supposed to cost.

Additions

A change order request should be in writing, stating exactly what is needed. It's a good idea to include drawings or sketches along with the detailed change instructions. Keep a careful record of change requests and drawings for each job.

The written notice may include a change in the time needed to complete the contract. This can be very important, especially if the contract provides liquidated damages for delay. Some penalties can amount to hundreds or thousands of dollars for each day of delay.

No contractor wants liquidated damages deducted from his contract price. If your general contractor has to pay damages for your delay, he may pass the charges on to the sub responsible for holding up the job. Usually he'll pick those who have the largest subcontracts.

Protect yourself by submitting a written notice to the contractor when your work is delayed. Request additional time when your work has been changed by change order. The more documentation you have, the better your chances of beating a backcharge for liquidated damages.

A sample change order is shown in Figure 14-2. The change is a simple addition to the building.

Deductions

Sometime the change will mean that less work has to be done, resulting in a saving to the owner. The contract provision for change orders usually states that deductive (credit) change orders don't require adjustment to overhead and profit. That means your overhead and profit are the same after the change. Only labor and material costs are deducted from the contract price.

If the east addition to the Advertising Company building in the previous example was a *deductive* change order, the material pricing and labor units would be about the same for those items in the job estimate. See the deductive change order at the end of this chapter.

Note carefully any work that has been completed before the deductive change order is expected to be approved. Take that material and its labor cost off the pricing sheets for the change. The owner should pay for material already in place even if he changes the plan to delete those items. Often, conduit rough-in has already been started. Once used, the material can't be salvaged for reuse.

Check your fixture and panelboard order for possible cancellation or restocking charges. Adjust the credit price downward for any material that won't be used and must be removed from the job.

Inform field workers of the deductive change being considered so that they'll stop work in the area affected by the change.

Adjustments may be required in the remaining circuiting after deletion of work. Check with the installers to confirm accurate adjustments that may not be apparent on the drawings.

Associated Contractors, Inc.
707 Washington Street
Industrial Park

Attention: Mr. Richard Giolando

Reference: Advertising Company
           East Addition
           Change Order Number 1

Gentlemen:

Transmitted herewith is a request for a cost breakdown for the East Addition to the referenced project. Appropriate X drawings are included with this request. All construction is to be in strict accordance with the Contract specifications and the special provisions of the Contract.

Submit a cost breakdown (6 copies) for each division of the work that is involved with this change request. The change order cost must be approved in writing before notice to proceed is granted.

The change consists of an addition 20' x 60' from column line 4 to new column line 5 and from column line D to column line G.

Cut three (3) 72" x 72" openings in the East wall of the Shop as indicated. Install approved roll-up steel doors with approved fire closing devices. The doors shall be capable of gravity closing in case of fire.

The construction of the addition shall be the same as was specified for the Shop area. See Shop area details on the Contract drawings.

Sample Change Order
Figure 14-2

Section 16 Electrical

Install industrial type lighting fixtures as indicated. Install the lighting on continuous strut channel and suspend with number 1/0 steel jack chain and appropriate fittings.

Extend the sound system to each section of the addition. Connect to the existing sound system in the Shop area. Use the same type and manufacturer of the approved speakers.

Install a new 100A, 120/208 Volt, 3 Phase, 4 Wire, 12 circuit panelboard on the North wall of the East addition. The panelboard shall be the same type and manufacturer as was approved for this Contract. Extend the existing underground 1-1/2" conduit to the new panel and install 4 #2 THW Cu cables and connect to the 100A spare circuit breaker in the Main Switchboard. Change the nameplate from spare to "Panel E."

Install a combination switch and receptacle in each location as shown. The switch shall control all of the lighting in the section.

Extend the exterior flood lighting to the new location and add the required number of flood lights as indicated. The new fixtures shall be the same type and manufacturer as was approved for the Contract.

Sample Change Order
Figure 14-2 (continued)

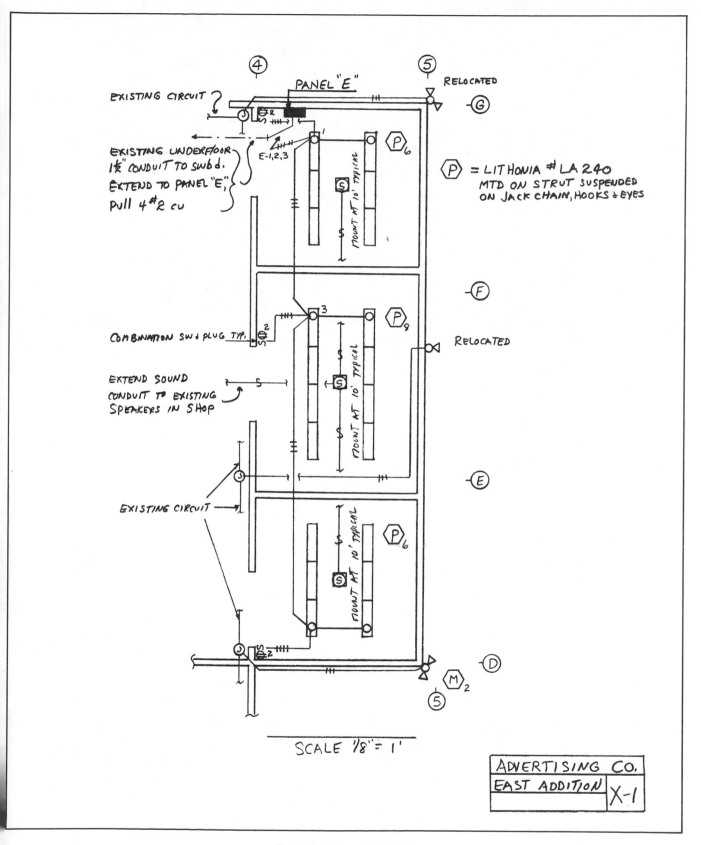

Sample Change Order
Figure 14-2 (continued)

MOUNTING __SURFACE__   PANEL __E__   MAIN __M.L.O.__

__120/208__ VOLTS __3__ PHASE __4__ WIRE   BUS __100A__

#	Description	A	B	C				A	B	C	Description	#
1	STORAGE LTG	1200			20		20	300			STORAGE PLUGS	2
3	STORAGE LTG		800		20		20		1000		SPARE	4
5	SPARE			1000	20		20			1000		6
7			1000		20		20	1000				8
9			1000		20		20		1000			10
11				1000	20		20			1000		12

WATTS	2200	1800	2000		1300	2000	2000

TOTAL WATTS   11300 = 11.3 KW =   33 AMPS

3762 X .25 ÷ 360 =   LCL   2.6 AMPS

ADVERTISING CO	
EAST ADDITION	X-2

Sample Change Order
Figure 14-2 (continued)

Cost Breakdown

Prepare the change order cost breakdown as accurately as possible. List every item of material on the pricing sheet. List all labor items and the cost of any special equipment that may be needed to make the change. (See Figure 14-3.) Carefully count each material item, and accurately measure all lengths of conduit and wire.

The design team will probably check the cost breakdown for accuracy and can request corrections or a resubmittal, if necessary. They may request a meeting to review apparent inaccuracies. If the breakdown is carefully prepared, clearly defined and mathematically correct, it should be difficult to dispute. A well-prepared cost breakdown is shown in Figure 14-4.

Usually, the person checking the change order pricing wants to reduce the total cost of the change. Calculations will be checked for extension or addition errors. Quantities may be checked for accuracy.

The following suggestions will help you prepare an accurate change order cost breakdown:

• Double check material and labor extensions and additions.

• Type the pricing sheets to ensure clarity.

• List only the total hours at the bottom of the pricing sheet. (Don't list individual labor units. Anyone can argue how much time it takes to install a specific item or material.)

• Prepare a simple but neat summary sheet.

• Check the contract for allowable overhead and profit markup.

• Clearly identify the change order.

• Above all, be prompt. Respond to change requests as quickly as possible. (See Figures 14-5 through 14-7.) Some contractors insist that the cost breakdown be presented on their forms.

When all materials and installations have been listed, write in the prices. Use standard list prices for all materials. In some cases, allowances must be made to offset small quantities and additional purchasing and handling costs. Freight must be considered even for small items that have to be shipped long distances.

Check with your distributor or supplier when returning on-site material for credit. Usually there's a substantial restocking charge, plus freight costs. There may be a cancellation charge for undelivered material. For example, lighting fixtures that were changed but not shipped may be subject to a cancellation charge.

Most manufacturers don't want material returned. Returned items can create unwanted surpluses. And usually the materials have been specially packed, labeled and marked for shipment to the job site. When an order is cancelled, the supplier has to unpack and restock the items. This can involve a great deal of processing and handling. The result is that you may be charged for the cancellation.

Always prepare the change order cost breakdown with good solid backup information. Use established pricing guides that include material costs. When special new material is needed, get a written quotation. Always be able to justify your prices in a change order.

Use the original estimate pricing sheets when setting credit prices. Be sure to discount some of those prices for the losses that will be incurred.

Review Meetings

The design team may not be satisfied with the change order cost breakdown submitted for their approval. Usually a meeting is called to resolve any problems. Some government agencies require that change orders be reviewed in a meeting with the contractor. For large changes many agencies require an audit of the cost breakdown.

When you attend a review meeting, be ready to document each material and work item. The breakdown must be neat, orderly and easy to understand. Those reviewing the breakdown want to reduce costs as much as possible. Good backup data is hard to argue with. Vague or poorly documented items will probably be rejected.

You don't submit change order costs on a take-it-or-leave-it basis. They're subject to full review and verification. Always keep your backup data and notes with the cost breakdown for future use.

At the change order review meeting, keep an open mind. Poor communication may have led to misunderstanding about the change in the first place. The designer may have meant one thing and you interpreted it to mean another.

Work Sheet

Estimate No.: __M-44__

ADVERTISING CO.

	CHANGE ORDER X	X-1
PANEL-E	1	
FIXT. M	2	
" P	20	
250 HPS	2	
F 40	40	
S	3	
S (in box)	3	
1/2" EMT	300	
1/2" CONN	30	
1/2" COUPL	28	
1/2" STRAP	30	
1 1/2" PVC	10	
1 1/2 Elb.	1	
1 1/2" FA	1	
1 1/2 RIGID	10	
1 1/2 TERM	1	
#12	750	
#2	260	
4/S BOX	12	
4/S BLANK	9	
4/S COMBO INDUST.	3	
1/0 CHAIN	100	
EYE SCREWS	12	
1/4" EYE BOLTS	12	
P1000 UNISTRUT	80	
1/4 SP NUTS	52	
B/C #14 SO CORD	100	
CORD CONN	12	
RELOCK TE FLOOD	3	
3/8 ANCHORS	4	
NAMEPLATE	1	
(1) WP	1	

Figure 14-3

Pricing Sheet

Job: ADVERTISING **Estimate No.:** M-44
Work: CHANGE ORDER NO.1 **Sheet:** 1 of 2
Estimator: ED **Checker** LUCY **Date:**

Description	Qty.	Price	Per	Extension	Hours	Per	Extension
PANEL "E"	1	—		87.00	—		3.50
FIXTURE M	2	296.00	E	592.00	1.00	E	2.00
" P	20	51.40	E	1028.00	.75	E	15.00
LAMPS 250 HPS	2	65.75	E	131.50	.10	E	.20
" F40 CW	40	2.50	E	100.00	.10	E	4.00
1/2" EMT CONDUIT	300	21.78	C	65.34	3.50	C	10.50
1/2" CONNECTORS	30	37.50	C	11.25	.08	E	2.40
1/2" COUPLINGS	28	48.80	C	13.66	.08	E	2.24
1/2" STRAPS	30	12.70	C	3.81	.08	E	2.40
1½" PVC CONDUIT	10	52.45	C	5.25	4.00	C	.40
1½" ELBOW	1	—		1.66	—	—	.15
1½" FA	1	—		.61	—	—	.15
1½" RIGID CONDUIT	10	79.33	C	7.93	7.00	C	.70
1½" LOCKNUTS	2	.33	E	.66	.10	E	.20
1½" INSUL. GR. BUSHING	1	—		2.44	—	—	.15
4/S BOX	12	117.45	C	14.09	.30	E	3.60
4/S COMBO INDUST.	3	98.75	C	2.96	.10	E	.30
4/S BLANK	9	43.90	C	3.95	.10	E	.90
#12 THHN CU	750	74.70	M	56.03	8.00	M	6.00
#8 TW CU	70	210.00	M	14.70	11.00	M	.77
#2 THW CU	260	662.00	M	172.12	17.00	M	4.42
3/C #14 SO CORD	100	850.00	M	85.00	15.00	M	1.50
S 1221-1	3	906.50	C	27.20	.35	E	1.05
⊖ 5242-1	3	256.00	C	7.68	.35	E	1.05
Ⓢ SPEAKER	3	64.00	E	192.00	.50	E	1.50
1/0 JACK CHAIN	100	20.00	C	20.00	10.00	C	10.00
1/4" EYE BOLTS	12	22.16	C	2.66	.10	E	1.20
1/4" EYE SCREWS	12	22.33	C	2.68	.10	E	1.20
Total				2652.18			17.48

Cost Breakdown
Figure 14-4

Pricing Sheet

Job: __ADVERTISING CO.__ Estimate No.: __M-44__

Work: __CHANGE ORDER NO. 1__ Sheet: __2__ of __2__

Estimator: __ED__ Checker __LUCY__ Date: _____

Description	Qty.	Price	Per	Extension	Hours	Per	Extension
P1000 UNISTRUT	80	217.30	C	173.84	8.00	C	6.40
P1006-1/4 20 NUTS	52	57.20	C	29.74	.05	E	2.60
CORD CONN. 2 SCREW	12	.46	E	5.52	.10	E	1.20
RELOCATE FLOODS	3	—	—	—	1.50	E	4.50
3/8 ANCHORS/NUTS	4	60.76	C	2.43	.20	E	.80
① WP	1	—	—	4.35	—	—	.50
NAMEPLATE	1	—	—	5.00	—	—	.30
SOUND WIRING	80	—	—	130.00	—	—	—
LOCATE U.G. STUB	1	—	—	—	—	—	.50
TRENCH 10x12	10	—	—	—	—	—	1.00
CORE WALL	2	50.00	E	100.00	.30	E	.60
MISCL. MATERIAL	LOT	—	—	10.00			.50
Total				460.88			18.90

Cost Breakdown
Figure 14-4 (continued)

Associated Contractors, Inc.   4/7/82
707 Washington Street
Industrial Park

Attention: Mr. Richard Giolando

Subject: Advertising Company
            Change Order #1, East Addition
            Cost Breakdown

Gentlemen:

    Submitted herewith is our cost breakdown for Change Order Number 1.
The breakdown lists all of the material and labor that will be required
under the Electrical Section of the change. All of the material is the same
quality and manufacturer as that which was formerly approved for the project.

    The total additional cost is Five Thousand, Eight Hundred, Fifty
Six Dollars $5,856.00

    Please advise acceptance as soon as practical so that we can proceed
with the change and the ordering of the materials.

                            Very truly yours,

                            Ed

Change Order Cost Breakdown
Figure 14-5

Advertising Company   Date   7/20/82

Change Order Number 1, East Addition

Job Number M-44

_____

1. Material   3,113.06

   Tax @3%   93.39

2. Labor - 96 Hours @ 10.40/E   998.40

   Supervision @ 15%   149.76

3. Equipment, Scaffold   275.00

4. Subtotal   4,629.61

5. Overhead @ 15%   694.44

6. Net Cost   5,324.05

7. Profit @ 10%   532.41

8. Total ADDITIONAL Cost   5,856.00

Change Order Cost Breakdown
Figure 14-5 (continued)

Pricing Sheet

Job: Advertising Company   **Estimate No.:** M-44

Work: Change Order Number 1   **Sheet:** 1 of 2

Estimator: Ed   **Checker** Geana   **Date:** _____

Description	Qty.	Price	Per	Extension	Hours	Per	Extension
Panel "E"	1	----	E	87.00			
Fixture "M"	2	296.00	E	592.00			
Fixture "P"	20	51.40	E	1028.00			
Lamps 250 HPS	2	65.75	E	131.50			
Lamps F40CW	40	2.50	E	100.00			
1/2" EMT Conduit	300	21.78	C	65.34			
1/2" EMT Connector	30	37.50	C	11.25			
1/2" EMT Coupling	28	48.80	C	13.66			
1/2" EMT Straps	30	12.70	C	3.81			
1-1/2"PVC Conduit	10	52.45	C	5.25			
1-1/2" PVC Elbow	1	----	-	1.66			
1-1/2" PVC FA	1	----	-	.61			
1-1/2" Rigid Conduit	10	79.33	C	7.93			
1-1/2" Locknuts	2	.33	E	.66			
1-1/2" Insul. Gr. B	1	----	-	2.44			
4/S Box	12	117.45	C	14.09			
4/S Combo Cover	3	98.75	C	2.96			
4/S Blank Cover	9	43.90	C	3.95			
#12 THHN Cu	750	74.70	M	56.03			
#8 TW Cu Green	70	210.00	M	14.70			
#2 THW Cu	260	662.00	M	172.12			
3/C #14 SO Cord	100	850.00	M	85.00			
S 1221-I	3	906.50	C	27.20			
Duplex 5242-I	3	256.00	C	7.68			
Speaker	3	64.00	E	192.00			
1/O Jack Chain	100	20.00	C	20.00			
Eye Screw 1/4"	12	22.16	C	2.66			
Eye Bolt 1/4"	12	22.33	C	2.68			
Total				2652.18			77.48

Change Order Cost Breakdown
Figure 14-5 (continued)

Pricing Sheet

Job: Advertising Company

Estimate No.: M-44

Work: Change Order Number 1

Sheet: 2 of 2

Estimator: Ed

Checker: Geana

Date:

Description	Qty.	Price	Per	Extension	Hours	Per	Extension
P1000 Unistrut	80	217.30	C	173.84			
P1006-1420 Nuts	52	57.20	C	29.74			
Cord Connectors	12	.46	E	5.52			
Relocate Floods	3	----	–	----			
3/8" Anchors/Nuts	4	60.76	C	2.43			
W.P. J-Box	1	----	–	4.35			
Nameplate	1	----	–	5.00			
Sound Wiring	80'	----	–	130.00			
Locate U.G. Stub	1	----	–	----			
Trench 10 x 12	10	----	–	----			
Core Wall	2	50.00	E	100.00			
Miscl. Material	Lot	----	–	10.00			
Total				460.88			18.90

Change Order Cost Breakdown
Figure 14-5 (continued)

The owner's representative should head the meeting. Other trades may attend to review their cost breakdowns. Listen to the questions and answers, and be ready for the same kind of review of your breakdown.

Most of those at the meeting don't fully understand how the change will be made or what its effect will be on the work in place or on the rest of the system. If they don't understand these things, they probably won't approve your cost breakdown. Be as informative and helpful as possible. It's to your advantage that your change be clearly understood.

Be willing to consider adjustments to your change. Perhaps you've included something that isn't really necessary. Don't be stubborn. You're trying to clear up difficulties, not create them.

Be prepared to show where your unit prices came from. Have copies of current price lists on hand. Try to have written quotations for special equipment. Be sure that those at the meeting agree with the material lists, quantities, measurements, extensions and additions. Save the labor breakdown until last.

If necessary, show your original rough change order pricing sheets with the labor units for each item. If you have a published list of electrical labor units applicable to change orders, use it for reference. Generally, labor units should not exceed those standards.

Perhaps the owner's representative can't find any discrepancies in your cost breakdown but still insists that the price be reduced. Don't agree. Instead, offer to go back to your office and run through the change once more. Then show the contractor that you can't possibly reduce your change price. He may be able to convince the owner that your figures are accurate.

If that doesn't work and the contractor insists that you cooperate with the owner's representative, agree to the reduction under protest. Notify the contractor, in writing, that you will proceed as directed for the reduced price "under protest." Later, you may be able to resolve the difference in another change or by filing a claim against the project.

The Change Order Process
On a large project, a change order passes through many channels. This usually requires quite a bit of time for processing. The owner may authorize the change and then negotiate the price at a later date. This helps keep the project on schedule but may result in a substantially different change order price.

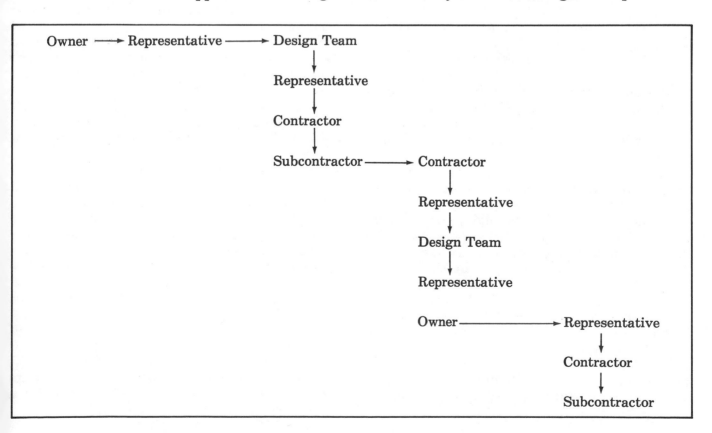

It can easily take a month to process a change order: from the time the subcontractor receives notification of the change until the owner issues the notice to proceed. The process usually involves the following steps:

1. The owner decides to ask for a change.
2. The representative checks the request and distributes it to the design team.
3. The design team works out the details of the change and returns their instructions and sketches to the representative.
4. The representative sends the request to the contractor.
5. The contractor sends the request to the subcontractors who will be affected.
6. The subcontractors prepare a cost breakdown and forward it to the contractor.
7. The contractor adds his portion of the change work, adds the markup, and sends the package to the representative.
8. The representative distributes the cost breakdowns to the design team members.
9. The design team returns their findings on pricing accuracy to the representative.
10. The representative reviews the findings and assembles the total change order pricing for the owner.
11. The owner decides if he still wants the change. The decision is returned to the representative.
12. The representative notifies the contractor.
13. The contractor notifies the subcontractor.
14. The owner authorizes a change in the contract price to cover the change order.
15. The contractor then issues a change in the subcontract price to cover the change order.

When the contract price has been revised to include the change cost, the change has been accepted by the owner. After acceptance, buy the material immediately and notify the installers. Be sure to include change costs on the progress billing invoices. Identify the change with the number assigned to it by the contractor.

Changes which have no effect on costs must also be accounted for and identified by the contractor. These need not be included on the progress invoices.

Subcontractor's Involvement

Most contracts want their subs to furnish cost breakdowns for change orders. Sometimes the contract simply states that the owner has the right to make additive or deductive changes.

When you get a change order request, do the following:

1. Review the total change request, even those sections that seem to be for other trades.
2. Review the original drawings and specifications, and identify the area where the change is to take place.
3. Check the total impact of the change request.
4. Make a detailed take-off of all material and work items.
5. Check the job site to verify the actual conditions, and make notes and take a few pictures, if necessary.
6. Adjust the take-off to reflect actual job site conditions.
7. Determine how far construction will have progressed by the time the change is approved.
8. Complete the pricing and submit the required copies of the cost breakdown.

Contractor Involvement

All change orders have to be processed through the general contractor. If the owner comes to you with a change, even a minor one, notify the contractor. Process the change through him.

Remember, your contract is with the contractor. The contract provides for payments to be made by the contractor for your completed work. You don't have a contract with anyone else, including the owner. The contractor won't honor invoices for payment of work done outside of his contract. If the change price hasn't been added to your contract, it doesn't exist. When changes involving your work are sent through the contractor, he can apply the agreed markup for his overhead and profit.

If the owner asks you to make a change, jot down the change request and notify the contractor. Then prepare a cost breakdown for the change and send it directly to the contractor for approval. He'll notify you if the change has been approved. Then you can begin the change work.

Sometimes you have to make a change yourself because of obvious design errors. First, notify the contractor of the problem. Let him know that you're preparing a cost breakdown for the change. Submit the cost breakdown as soon as possible and wait for approval. Don't submit the cost breakdown without first notifying the contractor. And certainly don't make the change and then try to get paid. The owner will withhold approval and stall until it's too late to remove the changed work.

Any change order sent directly to you from the owner or his representative should be turned over immediately to the contractor. Then wait for the contractor's instructions before making the change.

After the contractor receives the change from the owner, he'll issue change orders to the subcontractors who submitted approved cost breakdowns. The change order should show any change in the subcontract price and construction time.

A contractor can issue change orders to a subcontractor without involving the owner or his representative. An example would be a change order to cover the cost of installing temporary electric service equipment at the job site. These changes are strictly between the contractor and the subcontractor. Extensions to construction time are never considered in these change orders.

Effect of Change Orders

A change order will usually slow down progress on a job. Many tradesmen don't like to change installed work. They don't enjoy removing, replacing or altering work that has taken time and care to install. Consequently, changes usually take much longer than the original installations.

Most projects are stocked with bulk quantities of material ordered from the take-off lists. Deliveries are scheduled to coincide with the construction schedule. Panels, lighting fixtures and other special items are purchased and released to fit that schedule. Labor is supplied to maintain the construction schedule.

The supply of on-site material may have to be changed. New materials may be required. Others may no longer be necessary. Some may have to be removed or scrapped. Standing purchases may have to be changed.

Special materials: Changes in special material items, such as panelboards, can be very costly. Panelboards are usually built to job requirements. Circuit breaker arrangements may be custom made to the designer's panel schedule. The cover usually has a nameplate that identifies the panel.

Lighting fixtures are purchased according to job requirements. They're specially packaged and labeled for shipment to the project. Manufacturers don't want material returned to the factory. Most do not maintain stock for future sales. Some suppliers will take the material back only at a greatly reduced price, hoping to sell it at list price later.

Surplus materials: Unused materials such as conduit, fittings, boxes, wire, and devices are usually stored at the job site in makeshift areas. By the time the surplus is moved back to the contractor's yard, it no longer looks new. Often, it can't be used on other jobs. Broken bundles of conduit and opened boxes of wire are seldom permitted on most projects. The surplus may be a total loss. And the additional labor used for loading, hauling, stocking and taking inventory is a total loss.

When the project is finished and the surplus material is boxed up, certain items may no longer be complete. Special mounting screws, ballasts, sockets or other parts may have been borrowed from one item to install on another. The material left over is no longer complete. Often, it must be junked.

The credit given for surplus material on a deductive change order will probably be much less than the original cost. Sometimes, you'll be lucky to get anything.

Additional materials: When the change requires small amounts of additional material, adjust the costs for the *will-call* prices. These are usually higher than the original material, which is shipped directly to the job. Smaller quantities are usually close to list price.

There are additional handling costs for the will-call items. New items often must be special ordered. These shipping costs are added to the list price.

Suppliers usually don't stock special materials. But they will backorder any materials you need. Keep track of all backordered material. Rush it to the job site as soon as it arrives.

Additional labor: Additional labor is necessary for ordering and receiving new or special material. Time is needed to lay out the changed work. The normal routine is interrupted, so the scheduled work will be affected.

Often, new work is delayed because another trade required for the job is not on site. They may have finished their installation and are now working on another project, or they may not be scheduled to start on your project for another few weeks. All this can add up to lost time.

Too many minor changes can be confusing. Keep track of each minor change. If a change is not made at the right time, other work may have to be removed to accommodate it.

No Change Advantage

Many owners and their representatives feel that contractors take advantage of them when they request a change. Most don't realize the amount of disorder a change can create or the effect it can have on construction progress.

A contractor can do a better, faster, more profitable job when there are no change orders. A "no change" job is well designed and properly scheduled. There's little or no interference from the owner throughout the duration of the job.

At times, you'll want to make slight adjustments at no change in cost. An additive or deductive change can be so minor that it would cost more to furnish a cost breakdown than it's worth. A lighting fixture or some other special material may be deleted from the project. Make a "no cost" change with the understanding that the owner will receive the surplus item.

Work for Other Subcontractors

Other subcontractors on the project may want you to do work for them. A request from the heating, ventilating and air conditioning subcontractor would be typical. He's responsible for all of the HVAC control wiring but may ask you to price the control work. If he approves of the price, he'll issue a purchase order.

Make a complete take-off to determine the costs involved. Check the job site to be sure that all of the required work is understood. Prepare the proposal as if it were for a new small job—if it's in the early stages. If construction is at the end of the rough-in phase, price the work like any other change order.

Claims

Some day you will have to file a claim on a change order, extra work not covered under the subcontract, or some other dispute that might surface on a job. Most subcontracts have provisions for claims and disputes. The following is a typical claim provision:

The Subcontractor shall make all claims promptly to the Contractor for additional costs, extensions of time and damages for delays or other causes in accordance with the Contract Documents. Any such claim which will affect or become part of a claim which the Contractor is required to make under the Contract Documents within a specified time period or in a specified manner shall be made in sufficient time to permit the Contractor to satisfy the requirements of the Contract Documents. Such claims shall be received by the Contractor not less than two working days preceding the time by which the Contractor's claim must be made. Failure of the Subcontractor to make such timely claim shall bind the Subcontractor to the same consequences as those to which the Contractor is bound.

Some subcontract documents include a clause for arbitration. The following is a typical arbitration clause:

All claims, disputes and other matters in question arising out of, or relating to, this Subcontract, or the breach thereof, shall be decided by arbitration, which shall be conducted in the same manner and under the same procedure as provided in the Contract Documents with respect to disputes between the Owner and the Contractor, except that a decision by the Architect shall not be a condition precedent to arbitration. If the Contract Documents do not provide for arbitration or fail to specify the manner and procedure for arbitration, it shall be conducted in accordance with the Construction Industry Arbitration Rules of the American Arbitration Association then existing unless the parties mutually agree otherwise.

Another subcontract may simply state the following, which is recommended by the American Arbitration Association:

Any controversy or claim arising out of or related to this Contract, or the breach thereof, shall be settled by arbitration in accordance with the Construction Industry Arbitration Rules of the American Arbitration Association, and judgment upon the award rendered by the Arbitrator(s) may be entered in any Court having jurisdiction thereof.

According to the American Arbitration Association, an agreement written into a construction contract makes it easier and quicker to arrive at a peaceful settlement without having to go through arbitration at all. Thus, the arbitration clause is a form of insurance against loss of goodwill.

Most contract provisions don't permit filing of claims after work has been accepted and paid for.

The government has its own provisions for claims. Figure 14-8 shows typical claim provisions found in most government contracts.

Subcontractor's Claims

Most standard subcontracts give the subcontractor the right to make claims. They usually

1. *This Contract is subject to the Contracts Disputes Act of 1978 (P.L. 95-563).*

2. *Except as provided in the Act, all disputes arising under or relating to this Contract shall be resolved in accordance with this clause.*

3. *(a) As used herein, "Claim" means a written demand or assertion by one of the parties seeking, as a matter of right, the payment of money, adjustment, or interpretation of Contract terms, or other relief, arising under or relating to this Contract. However, a written demand by the Contractor seeking payment of money in excess of $50,000 is not a claim until certified in accordance with 4, below.*
   (b) A voucher, invoice, or other routine request for payment that is not in dispute when submitted is not a claim for the purpose of the Act. However, where such submission is subsequently disputed either as to liability or amount or not acted upon in a reasonable time, it may be converted to a claim pursuant to the Act by complying with the submission and certification requirements of this clause.
   (c) A claim by the Contractor shall be made in writing and submitted to the Contracting Officer for decision. A claim by the Government against the Contractor shall be subject to a decision by the Contracting Officer.

4. *For Contractor claims of more than $50,000, the Contractor shall submit with the claim a certification that the claim is made in good faith; the supporting data are accurate and complete to the best of the Contractor's knowledge and belief; and the amount requested accurately reflects the Contract adjustment for which the Contractor believes the Government is liable. The certification shall be executed by the Contractor if an individual. When the Contractor is not an individual, the certification shall be executed by a senior company official in charge at the Contractor's plant or location involved, or by an officer or general partner of the Contractor having overall responsibility for the conduct of the Contractor's affairs.*

5. *For Contractor claims for $50,000 or less, the Contracting Officer must, if requested in writing by the Contractor, render a decision within 60 days of the request. For contractor claims in excess of $50,000, the Contracting Officer must decide the claim within 60 days or notify the Contractor of the date when the decision will be made.*

6. *The Contracting Officer's decision shall be final unless the Contractor appeals or files a suit as provided in the Act.*

7. *Interest on the amount found due on a contractor claim shall be paid from the date the Contracting Officer receives the claim, or from the date payment otherwise would be due, if such date is later, until the date of payment.*

8. *The Contractor shall proceed diligently with performance of this Contract, pending final resolution of any request for relief, claim, appeal or action arising under the Contract, and comply with any decision of the Contracting Officer. (DAR-103.12)*

Typical Claim Provisions in Government Contracts
Figure 14-8

state that the subcontractor must abide by all applicable provisions of the contract documents. These provisions usually require that the subcontractor submit all correspondence and claim data through the contractor, who then gives the submittal to the owner or his representative. The contractor usually adds the allowable markup for administering the addition to the contract as if it were a change order.

Preparing a Claim

Prepare the claim according to the procedures given in the contract provisions. If the contract doesn't state the procedures for claims, follow those established for change orders. If there aren't any, use standard company procedures for claims or change orders.

List all relevant costs as clearly as possible. Check and double check all extensions and additions. Prepare a report containing all facts relating to the claim or dispute. Assemble as much backup data as possible. Refer to job records, the job diary, dates, times, past correspondence, and notes taken at meetings. Include any other items you feel are important. Neatly assemble the claim or dispute and submit it promptly to the contractor.

Send your claim submittal by special messenger or by return receipt delivery service. Be sure to obtain a receipt from the contractor for the submittal.

Follow up within three or four days, in writing, asking what action is being taken. Follow up again on a regular basis at least once a week until you receive some acknowledgement.

Give copies of all correspondence to those in your company involved with the claim. List their names and titles on each copy. If you have an attorney on retainer, include the attorney's name also. Be sure that he receives copies.

If the matter must go to arbitration, have your attorney prepare and submit your request for arbitration to the American Arbitration Association in your area. Figure 14-9 lists AAA regional offices.

The best way to win your claim is to keep good job records. Keep copies of all correspondence, and follow up conversations regarding the job. Good, clear, factual records are hard to dispute.

A claim or dispute may go to arbitration or to court. If you've handled the matter to the best of your ability and your company has performed in good faith, you shouldn't have too much trouble.

Mechanic's Lien

A mechanic's lien is a charge placed on real property for satisfaction of unpaid debts for work done under contract. The claimant can force the sale of the property and obtain relief out of the proceeds of the sale.

The mechanic's lien doesn't necessarily provide an orderly cash flow, nor does it offer a fool-proof guarantee of full relief. The procedure for filing a lien is complicated and slow. It involves legal expenses, court costs and other costs which usually prevent full recovery. Other remedies are available, but most are just as complicated.

Stop Notice

A stop notice informs the person holding the construction funds to withhold money to pay the claimant for the work performed or for materials supplied. This notice, subject to certain qualifications, obligates the person holding the funds to retain an amount sufficient to pay the claimant. Also, it places a lien on the funds in favor of the claimant.

Payment Bond

A payment bond is a guarantee by an insurance company to one or more contractors that they will be paid for their work. This gives you the right to payment from the insurance company if you have not been paid by a contractor, owner or construction lender.

Stop notices and payment bonds are available on both public and private work, although a payment bond is not mandatory on private work. However, for reasons of public policy, a mechanic's lien can't be made on real property owned by local, state, or federal governments. Therefore, either a stop notice or a payment bond is used on public works projects. The Miller Act, a federal law, provides that the only remedy for work done on federal projects is suit on a mandatory payment bond.

If the person holding the construction funds doesn't honor the stop notice, personal liability may result. The mechanic's lien is merely a lien on the improvement and, to a limited extent, on the land on which the improvement was made. There is no personal liability connected with

American Arbitration Association

Regional Offices

Atlanta, 100 Peachtree Street, N.W. 30303

Boston, 294 Washington Street 02108

Charlotte, P.O. Box 18591, 3235 Eastway Drive 28218

Chicago, 180 N. La Salle Street 60601

Cincinnati, 2308 Care Tower 45202

Cleveland, 215 Euclid Avenue 44114

Dallas, 1607 Main Street 75201

Detroit, 1234 City National Bank Building 48226

Garden City, N.Y. 585 Stewart Avenue 11530

Hartford, 37 Lewis Street 06103

Los Angeles, 443 Shatto Place 90020

Miami, 2250 S.W. 3rd Avenue 33129

Minneapolis, 1001 Foshay Tower 55402

New Brunswick, N.J. 96 Bayard Street 08901

New York, 140 West 51st Street 10020

Philadelphia, 1520 Locust Street 19102

Phoenix, 222 North Central Avenue 85004

Pittsburgh, 221 Gateway Four 15222

San Diego, 530 Broadway 92101

San Francisco, 690 Market Street 94104

Seattle, 810 Third Avenue 98104

Syracuse, 731 James Street 13203

Washington, 1730 Rhode Island Avenue, N.W. 20036

White Plains, N.Y. 34 South Broadway 10601

Contact the AAA office in your area for details about the Construction Industry Arbitration Rules and how contract disputes are resolved under those rules.

AAA Regional Offices
Figure 14-9

the mechanic's lien unless the claimant can prove existence of a contract. In this case, liability stems from the contract, not from the mechanic's lien.

Backcharges

A backcharge is a claim to recover extra costs between the contractor and the subcontractors. Usually the charges are small and are made to encourage the subcontractor to furnish his own tools, equipment and labor.

Backcharges are common for cleanup at the job site when a sub leaves trash on the site rather than disposing of it. They can also be for the loan of tools, ladders and other equipment. Cleaning, repairing and disputes between subcontractors are covered, too. Some contractors post rate lists of their charges in the project field office and inform each subcontractor that charges will be made for the loan of any tools, equipment or labor.

Safety rules on construction sites must be strictly enforced. When an employee of a subcontractor shows up for work without a hardhat, he should either get one or leave the site. If the sub doesn't have extra hard-hats, the contractor may furnish one at the posted rate. The rate is usually high enough to encourage subcontractors to furnish their own.

Sometimes the contractor will supply concrete for underground conduit encasement or for lighting standard foundations. The backcharge will usually be at the same rate as the cost to the contractor. When he's making a sizeable pour, the contractor's cost per cubic yard of concrete is generally lower than the cost for a short load to the subcontractor.

General contractors don't want to be responsible for ordering and scheduling the subcontractor's materials at the same price he's paying. When the subcontractor uses a few cubic yards of concrete for underground encasement, it may be a richer mix than necessary. The contractor may be pouring structural concrete of 3000 pounds per square inch (psi), and the underground encasement mix can be as low as 1000 psi. The subcontractor will have to pay for the 3000 psi concrete and any truck delay or standby time.

Trenching is another item to consider. The contractor may be able to do the excavation for the building foundations and the trenching for the electrical and plumbing underground. It may be cheaper to pay the hourly rate for on-site trenching done by the contractor than to pay separately. When the amount of trenching is less than a day's work, it begins to make even more sense. Coordination with the contractor and the plumbing subcontractor is essential to avoid cross ditching.

Sub to Sub Charges

Use purchase orders when charging another subcontractor for work. The charge can be predetermined for large amounts of work. An open purchase order can be used for small amounts.

When you correct work damaged by another subcontractor and a dispute arises, submit the charges to the general contractor. He will backcharge the other subcontractor. This speeds repairs without holding up other work.

If the contractor or the owner damages your work, submit a claim to the contractor. A detailed report should accompany the claim, as well as an explanation of the charges.

If the equipment you install doesn't work, your supplier should stand behind what he has sold you. He then must turn to his supplier or the manufacturer to correct the matter. It's wise for you to delay payment on a portion of the bill until the equipment has been tested. That amount could be about as much as the retainage withheld by the owner from each progress payment.

When all or part of the electrical system fails to operate because of a design flaw, immediately notify the contractor with a written report, possibly recommending corrections. The contractor will forward the report to the architect, who will send it to the electrical engineer. The electrical engineer will check the problem and should design a change to solve the problem.

This will usually involve a change order. The claim should be presented just like any change order.

You can't backcharge the contractor, owner, electrical engineer or another subcontractor. The only way to pass on the charges is to get a purchase order for the work or to file a detailed claim.

Supplier

At times you'll have to submit backcharges to a supplier for material or equipment problems. This amounts to withholding payment to the supplier. The backcharge must have good supporting reasons and figures so the supplier can backcharge the manufacturer.

A supplier backcharge could be for repairing defective items at the job site. An example would be errors in the control wiring of a motor control center. Maybe the wiring doesn't meet project requirements and control diagrams. There's not much actual material involved, but the corrections require a great deal of labor. There may be a defective control device in the center that's difficult to locate or a wire that has broken inside the insulation jacket.

Another common problem is wiring for fluorescent lighting fixtures. The manufacturer may have installed the wrong ballast, and the fixtures were installed before the problem surfaced. Maybe the manufacturer hasn't supplied the fixtures wired for the switching features specified. Some lighting designs require certain lamps to light when a certain switch is turned on. In a large office area, for example, with rows of fluorescent lighting fixtures having four lamps in each, the switching usually requires the two inside lamps to be connected to one switch and the two outside lamps to be connected to another. If the fixtures are not wired correctly, the lighting will not operate the way the code requires.

Sometimes equipment is shipped before it was completed and tested at the factory. Jumpers, wiring, control devices and control fuses may be missing. It takes time to locate and replace missing items.

Damaged Items
Material and equipment damaged during delivery are not necessarily backcharge items. When damaged items show up at the job site, the shipment should either be refused or be noted on the freight bill as damaged. Damage in transit is the problem of the freight line. But you have to make the claim.

Most freight line inspectors will suggest trying to fix the damage. In most cases that's not possible. The damage will usually be obvious. Get a copy of the inspector's report and immediately notify the supplier. Send a copy of the report, and request instructions for making repairs.

If concealed damage is found later, contact the freight lines. Get an inspection report and send a copy of it to the supplier, with a note detailing the problem. Ask for instructions on making the repairs or installing a replacement.

When you need to file a claim against a freight line for damaged material or equip-

ment, have the supplier involved help you. Make sure the supplier understands the problem. Have him provide instructions to get the repairs done properly.

Freight claims can be sticky. There are many ways a freight line can dodge a claim: timing, backup inspections, signatures, dates, split responsibilities with other freight lines and packaging problems are a few. You must have accurate and timely records for them to even consider settlement.

Project Delays
When the job has been delayed beyond the completion date and liquidated damages are assessed by the owner against the contractor, the contractor will pass the charges on to the subcontractors. If some of your equipment delivery caused the delay, you may be backcharged.

If you're having a hard time with a manufacturer about deliveries, notify the contractor as early as possible. Maybe he and the owner can pressure the manufacturer. Always give written notification, and send a copy to the supplier so he too can join in. If you complete the job promptly and according to plan, many delivery problems will be eliminated. When the delay comes, the owner will be aware of the problem and may not assess liquidated damages. If you don't inform the contractor of the problem, you can expect a backcharge.

If the job has been delayed by others not under your control and the contractor backcharges your company for a share of the damages, you may have a tough fight. The contractor has probably figured how big a backcharge you'll take without argument. Legal services are expensive, and a dispute with the contractor may hold up other money that's due. The struggle may take years to resolve.

Contractors usually make backcharges at the very end of a job. By that time you probably don't owe the supplier enough to cover any of the charges and probably can't find a reason to backcharge the supplier anyway.

Backcharge Effects
If backcharges become a continuing nuisance, have a talk with the contractor. Maybe you can work things out. If you can't, then don't bid to him again.

The same thing can happen to your accounts

with suppliers. They won't continue to take backcharges and still give you good service.

When the supplier receives backcharges, he must backcharge the manufacturers. Like anyone else, the manufacturer doesn't like backcharges. He'll drop a distributor who makes too many backcharges.

Good working relationships are the key to successful ventures. They take years to develop, but only days to ruin. If your treatment of contractors and suppliers is honest and sincere, they'll respond in the same way. Backcharges are hard to take. They leave everyone with bad feelings.

Associated Contractors, Inc.   7/20/82
707 Washington Street
Industrial Park

Attention: Mr. Richard Giolando

Reference: Advertising Company
           Change Order Number 1
           Delete East Storage Area

Gentlemen:

    Transmitted herewith is a request for the deletion of the Shop Storage Areas at the referenced project. A credit breakdown is required for all of the deleted work.

    All work shall proceed on the project, including the East Storage Area until approval of the credit, then that portion of the project shall be deleted.

    The deletion consists of elimination of the three (3) Storage Areas on the East side of the building between column lines 4 and 5 and between column lines D and G.

    Close the 8' x 8' opening in the East Wall with the same material as the wall. Provide a clean roof line all along the east wall.

    Place A/C pavement continuous along the East side of the building, same as detailed for parking area. Provide striping for parking places as is detailed for parking spaces.

    Relocate flood lighting to new locations, delete the two flood lights at column line D.

    Cap-off Panel E feeder conduit below grade two (2) feet outside the building line. Delete cables for Panel E. Delete the 100A circuit breaker in the Main Switchboard for Panel E. Provide a nameplate marked "Spare."

    Delete the sound system to the Storage Area.

    Provide a complete credit breakdown for all of the deleted work as soon as possible.

Deductive Change Order Request
Figure 14-10

Pricing Sheet

Job: **ADVERTISING COMPANY**   Estimate No.: **M-44**

Work: **DELETE STORAGE AREAS C.O. #1**   Sheet: **1** of **2**

Estimator: **ED**   Checker **SUE ✓**   Date: _____

Description	Qty.	Price	Per	Extension	Hours	Per	Extension
PANEL "E" LESS RESTOCK FRT	1			19.80			3.50
FIXT. TYPE "M"	2	138.00	E	276.00 ✓	1.00	E	2.00 ✓
" " "P"	20	11.75	E	235.00 ✓	.75	E	15.00 ✓
LAMPS 250 HPS	2	41.42	E	82.84 ✓	.10	E	.20 ✓
S F40 WWSS	40	2.36	E	94.40 ✓	.05	E	2.00 ✓
½" EMT CONDUIT	300	11.25	C	33.75 ✓	3.25	C	9.75 ✓
1½"	10	41.81	C	4.18 ✓	5.75	C	.58 ✓
½" CONNECTOR	30	22.00	C	6.60 ✓	.05	C	1.50 ✓
1½"	2	1.50	E	3.00 ✓	.10	E	.20 ✓
½" COUPLING	28	28.65	C	8.02 ✓	.05	E	1.40 ✓
1½" PVC CONDUIT	10	32.20	C	3.22 ✓	3.45	C	.35 ✓
1½" ELBOW	1	—		1.52	—		.10
1½" FA	1	—		.56	—		.10
4/S BOX	12	85.80	C	10.30 ✓	.20	E	2.40 ✓
4/S COMBO COVER	3	.70	E	2.10 ✓	.10	E	.30 ✓
4/S BLANK COVER	9	.33	E	2.97 ✓	.05	E	.45 ✓
#12 THHN CU	750	39.80	M	29.85 ✓	6.50	M	4.88 ✓
#8 THW	70	117.92	M	8.25 ✓	9.50	M	.67 ✓
#2 THW	260	360.00	M	93.60 ✓	15.00	M	3.90 ✓
3/C #14 SO CORD	100	595.00	M	59.50 ✓	15.00	M	1.50 ✓
S 1221-1	3	906.50	C	27.20 ✓	.30	E	.90 ✓
DUPLEX 5242-1	3	182.35	C	5.47 ✓	.30	E	.90 ✓
SPEAKER	3	—		300.00	—		FBO
SOUND WIRING	LOT	—		—			—
1/0 JACK CHAIN	100	16.00	C	16.00 ✓	10.00	C	10.00 ✓
EYE SCREW ¼"	12	17.86	C	2.14 ✓	.10	E	1.20 ✓
EYE BOLT ¼"	12	14.66	C	1.76 ✓	.10	E	1.20 ✓
P1000 UNISTRUT	80	197.50	C	158.00 ✓	7.00	C	5.60 ✓
Total				1546.03 ✓			70.58

Deductive Change Order Work Sheet
Figure 14-11

Pricing Sheet

Job: ADVERTISING COMPANY   **Estimate No.:** M-44

Work: DELETE STORAGE AREA   C.O. #1   **Sheet:** 2 of 2

Estimator: ED   **Checker** SUE ✓   **Date:** _____

Description	Qty.	Price	Per	Extension	Hours	Per	Extension
P1006-1420 NUTS	52	52.00	C	27.04 ✓	.05	E	2.60 ✓
CORD CONNECTORS	12	.33	E	3.96 ✓	.10	E	1.20 ✓
3⁄8" ANCHORS/NUTS	4	41.63	C	1.67 ✓	.20	E	.80 ✓
W.P. J-BOX	1	—		4.35			.50
TRENCH 10x12	10	—		—			1.00
100A 3P CIRCUIT BKR	1	—		78.00			—
				115.02			6.10 ✓
ADDITIONAL WORK:							
NAMEPLATE	1	—		5.00	—		.30
U.G. CONDUIT MARKER	1	—		10.00	—		1.00
1½" PVC CAP	1	—		.89	—		.10
C/B CLOSURE PLATE	1	—		30.00	—		.50
				−45.89 ✓			−1.90 ✓
	Total			69.13 ✓			4.20 ✓

Deductive Change Order Work Sheet

Figure 14-11 (continued)

Associated Contractors, Inc.   8/11/82
707 Washington Street
Industrial Park

Attention: Mr. Richard Giolando

Subject: Advertising Company
          Change Order Number 1, Delete Storage Area
          Credit Breakdown

Gentlemen:

     Submitted herewith is our credit breakdown for Change Order Number 1.
The breakdown lists all of the material and labor that is being credited back
from the electrical work that is not under Contract. The material and labor
costs are taken directly from our original cost estimate for the project.

     The panelboards are at the jobsite, so panel "E" will be returned
to our supplier. The lighting fixtures are being cancelled from our Purchase
Order to our supplier. The supplier will charge a restocking charge for the
returned panel and will charge a cancellation charge for the fixtures.

     The total amount of credit is Two Thousand, Five Hundred, Sixty
One Dollars $2,561.00

     Please advise acceptance as soon as practical so that we can instruct
our crew to delete the Storage Area work.

                              Very truly yours,

                              Ed

Deductive Change Order Cover Letter
Figure 14-12

Advertising Company   Date   7/20/80

Change Order Number 1, Delete Storage Areas

Job Number M-44

---

1. Material   1,615.16
   Tax @ 3%   48.45

2. Labor - 75 Hours @ 10.40/E   780.00
   Supervision @ 15%   117.00

3. Total Amount of Credit   2,561.00

Deductive Change Order Cost Breakdown
Figure 14-13

Pricing Sheet

Job: Advertising Company   **Estimate No.:** M-44

Work: Delete Storage Areas, C.O. 1   **Sheet:**____of____

Estimator: Ed   **Checker** Sue   **Date:** _____

Description	Qty.	Price	Per	Extension	Hours	Per	Extension
Panel "E"	1			79.80			
Fixt. Type "M"	2	138.00	E	276.00			
Fixt. Type "P"	20	11.75	E	235.00			
Lamps 250HPS	2	41.42	E	82.84			
Lamps F40CW-SS	40	2.36	E	94.40			
1/2" EMT Conduit	300	11.25	C	33.75			
1-1/2" EMT Conduit	10	41.81	C	4.18			
1/2" EMT Connector	30	22.00	C	6.60			
1-1/2" EMT Connector	2	1.50	E	3.00			
1/2" Coupling	28	28.65	C	8.02			
1-1/2" PVC Conduit	10	32.20	C	3.22			
1-1/2" PVC Elbow	1			1.52			
1-1/2" PVC FA	1			.56			
4/S Box	12	85.80	C	10.30			
4/S Combo Plate	3	.70	E	2.10			
4/S Blank Cover	9	.33	E	2.97			
#12 THHN Cu	750	39.80	M	29.85			
#8 THW Cu	70	117.92	M	8.25			
#2 THW Cu	260	360.00	M	93.60			
3/C #14 SO Cord	100	595.00	M	59.50			
S 1221-I	3	906.50	C	27.20			
Duplex 5242-I	3	182.35	C	5.47			
Speaker Box	3			300.00			
Sound Wiring	Lot			Included			
1/0 Jack Chain	100	16.00	C	16.00			
Eye Screw 1/4"	12	17.86	C	2.14			
Eye Bolt 1/4"	12	14.66	C	1.76			
P1000 Unistrut	80	197.50	C	158.00			
P1006-1420	52	52.00	C	27.04			
Total				1573.07			73.18

Deductive Change Order Pricing
Figure 14-14

Pricing Sheet

Job: Advertising Company   **Estimate No.:** M-44
Work: Delete Storage Areas, C.O. 1   **Sheet:** ___ of ___
Estimator: Ed   **Checker** Sue   **Date:** _____

Description	Qty.	Price	Per	Extension	Hours	Per	Extension
Cord Connectors	12	.33	E	3.96			
3/8" Anchors/Nuts	4	41.63	C	1.67			
W.P. J-Box	1			4.35			
Trench 10 x 12	10			----			
100A, 3P Cir. Bkr.	1			78.00			
				87.98			3.50
Additional Work							
Nameplate	1			5.00			
U.G. Stub Marker	1			10.00			
1-1/2" PVC Cap	1			.89			
C/B Close Plate	1			30.00			
				45.89			1.90
Total				42.09			1.60

Deductive Change Order Pricing
Figure 14-24 (continued)

Chapter 15

Cost Adjustments

Alternates

Some contracts permit the contractor to submit alternate proposals with the bid. These may include alternate construction methods, materials and equipment. Some bid forms specify that alternate bids be offered at the time of the bid. It's a wise owner that is willing to study more economical ways to build his project.

Most contracts specify certain types of material or equipment. It's common to see a specific catalog number in the specs. But it's just as common to see the words "or equal" following that number. This means that if the specified material can't be obtained, a suitable alternate may be substituted. But it's hard to determine at the time of bid whether an alternate is "equal" to the specified item.

Many suppliers have difficulty providing quotations for proprietary material and equipment. A proprietary item is usually under the exclusive control of a dealership or distributorship. The supplier must find an alternate that's equal to the one specified.

For example, consider the lighting fixtures for a medium to large building. The lighting fixture schedule lists dozens of types to be installed. Many are listed by manufacturer, catalog number and specific features. Very few suppliers could quote the entire list as specified. Usually they'll quote the fixtures they normally handle and quote alternates for those they don't.

Some suppliers may offer alternates with much lower prices to reduce overall costs. And some contractors look for alternates to lower their costs, also.

Some manufacturers ask higher prices for products to be used in special installations. They're gambling that the competition won't bid against them. If the competition does bid, the manufacturer may come back after the bid date and try to sell the winning contractor on his product.

In most cases, the contractor must show that the alternate is an acceptable substitute. Its acceptance depends on how strict the specifications are and how much the owner wants to maintain the quality of the material.

Bid Form Modifications

Most bidding documents prohibit modifications to the bid form. If the form calls for a lump sum price and a contractor decides to list some deductive alternatives or takes exception to any part of the bidding or contract documents, the bid will be considered non-responsive and will be ignored.

Planned Alternates

The architect may request certain planned alternates. Several portions of the building may be priced separately so that construction can be planned to stay within a construction budget. The contract award is usually made to the lowest bidder on the combination of alternates selected.

The alternates can be additive or deductive, depending on the bidding documents. *Additive* alternates are added to the basic bid. *Deductive* alternates are subtracted from the total bid. It's possible to have both on the same bid form.

Another planned alternate is unit cost items for minor adjustments during construction. The bidding documents may call for certain items to be priced separately. These might include the cost for adding a single-pole switch, a duplex receptacle, or some other minor item.

Bid alternates can be tricky. Consider them carefully and understand the scope of each before finalizing the bid.

Some contractors lower their basic bid a little, then add bid alternates. They figure that after the contract is let they can pick up the alternate work by change order. Sometimes it works out that way. The owner may be cautious when preparing the contract, but after a while he may be more receptive to adding certain alternate work.

Most contracts have provisions for adding selected alternate work at a later date. These provisions should state that the owner has the right to select any bid alternate at the price listed at the time of the bid. They should also mention that the alternate work won't affect work in progress.

If the contract doesn't have such provisions and alternates are not accepted, a change order is necessary to place the alternate in the contract. The price will probably be increased because of the impact the change will have on the project. The contractor would price alternate work about the same as basic work, knowing that the alternate would have little effect at the time the contract is executed.

Material and Equipment

Specifications usually set quality standards for materials and equipment. The contract documents probably state that substitute material and equipment won't be accepted unless approved in writing prior to the bid. After the bids are opened, no other alternate or substitution will be permitted. It would be very difficult to substitute an alternate under these conditions, especially when the bid is on a private project.

The specifications for most public works projects do not list proprietary material and equipment. Public jobs control quality by listing:

• At least three manufacturers on each callout.

• One manufacturer to establish a quality standard, and the designation of "or equal" immediately following the callout.

• A complete description of the item, including sketches and drawings. Manufacturer and catalog number are not mentioned.

The specifications may identify the manufacturer of equipment that must be matched. They may also give a complete description of the item. Check the specified item to determine exactly what is required for the alternate.

Reducing Costs

Construction funds on a project may be so limited that several alternate prices must be submitted with the base bid price. The owner usually wants all bids to be complete so the job can be scheduled properly. Delaying any part of the project might cause scheduling and coordinating problems. Adding the deleted portions after the contract has been completed is costly and would interfere with other work still in progress.

The owner may try to negotiate a contract with the low bidder. For example, the owner may ask the contractor to submit cost saving suggestions after the job is awarded. This gives the owner and contractor some flexibility. The owner might select several of the proposed cost reductions and reject others.

To meet his construction budget, the contractor may ask the major subcontractors bidding the job to find ways of cutting their costs. Many respond by substituting alternate materials of lesser quality than those in the specifications. You may be able to offer cost saving suggestions such as these:

• Reduce lighting fixtures to lesser quality alternates.

• Reduce special equipment to lesser quality alternates.

• Substitute loadcenters for panelboards.

• Use aluminum wire instead of copper feeder wire.

• Reduce wiring methods to standard code requirements.

• Reduce wiring devices to standard code requirements.

• Shift trenching to the contractor.

• Shift placement of concrete light standard bases to the contractor.

You can add many more items to this list. But make substitutions carefully and wisely. You don't want to reduce the quality of the overall installation.

Electrical work is usually about ten percent of the total bid price. Cost shaving on electrical work will seldom come to more than one percent of the bid price.

Don't be too eager to reduce the quality of the electrical installation. Many existing systems need upgrading just to keep up with technical advances. And upgrading a lower grade electrical system can be extremely expensive. Most of the system is concealed in walls, under floors, in ceiling systems and in special chases. Most loadcenters are difficult to change without shutting off the system in the service area.

Lighting fixtures that look similar may have very different lighting levels. And lower quality equipment usually has a lower useful life expectancy.

It's all right to use alternates to reduce costs, but don't go overboard. Let care and caution be your guide.

Re-engineered Estimates

Some electrical contractors provide electrical engineering services as part of their bid. Many specialize in supermarkets, industrial buildings, office buildings, churches, homes, schools, condominiums or mobile home parks. Those with a good reputation for doing quality jobs at a fair price have little trouble getting work.

Some electrical contractors select certain projects with the intention of completely re-engineering the work. They estimate the job as specified and make a competitive bid. If awarded the subcontract, they immediately re-engineer the electrical work to reduce installation cost, often by using less expensive materials and procedures. This is fine if the re-engineered system will provide the service called for in the plans and specifications. But it can't reduce the quality of the installation to the point where it is faulty or hazardous.

Here's how to re-engineer a job to cut costs:

• Reduce material quality to code grade standards.

• Combine circuits into common conduit runs.

• Load lighting and plug circuits to the capacity of the circuit breaker protecting the wire.

• Relocate panels to more central locations to cut down on the lengths of conduit runs.

• Reduce the size of panel and switchboards to the lowest capacity requirements.

• Eliminate special installation details and instructions wherever possible.

• Reduce wiring methods to minimum code standards.

Electrical contractors who specialize in re-engineering can bid jobs at a lower price. They know which jobs can be re-engineered and which can't. They tend to avoid jobs that require certain materials or approval of equipment shop drawings. These jobs usually require that all changes be approved before purchase and installation of materials.

When the electrical engineer or designer demands full compliance with the specifications or when a full-time inspector is on the job, the re-engineer specialist may find it difficult to rework a system. The electrical engineer may reject low quality alternates. The resident inspector will check all material and equipment to make sure it complies with the approved submittals. He'll also make sure that all installations are done according to instructions on the plans and in the specifications. Few resident inspectors are experts in all types of electrical work. But most know how to read and follow written specifications. If they come across something they don't understand, they'll probably consult the electrical engineer.

The resident inspector usually has a lot of control over the project. Often, he'll advise the owner on percentages of completion for the monthly progress payments. His job is to make sure that the owner gets what he contracted for.

Most resident inspectors have a good working knowledge of the overall job requirements. If there's a problem, the inspector can call in the architect or one of the division consultants. The problem may be in design or an installation method. The architect will study the problem and issue corrective action, if necessary. If the contractor is installing material that doesn't comply with the contract documents, he may have to replace that material.

On private work, local authorities will inspect the project. But they're only checking code compliance, a fairly minimum standard. The electrical contractor can reduce material or equipment quality and still get by the local authorities. But he'll have a harder time getting by the resident inspector.

If there is no resident inspector on the job, you may be tempted to try some re-engineering. Even when the architect and the project consultants make periodic inspections, work that's concealed is difficult to inspect.

Correcting the Changes

When the owner finds changes, it's usually too late to correct them. The owner doesn't want delays. He's committed to completing the project by a certain date. Most projects get started later than scheduled, and there are always unavoidable delays.

If he finds unauthorized changes that lower the quality of the system, the owner may withhold funds to force the contractor into negotiating a settlement. Sometimes, corrections can be made. Construction should remain on schedule while the adjustments are negotiated and corrections are made.

The credit offered in these negotiations is usually just a token amount and less than the true cost difference. But it does provide some relief and makes the contractor more aware of the job requirements—and less likely to make unauthorized changes.

Some changes may be extensive, requiring costly adjustments. The owner should have the work appraised by a third party experienced in cost estimating. The amount required to restore the work should be withheld from payment. If the contractor and the owner can't agree on the adjustments to be made, the contractor may have to settle the matter in court or in arbitration.

Finding Changes

On private jobs without a resident inspector, there are many opportunities to re-engineer the electrical design. The owner is usually too busy to be on the site at all times. The architect's

time is limited, and he's more concerned with the planned change orders and the overall esthetics of the project. The electrical consultant is much too busy to make periodic inspections and usually is not paid to do so.

During the final inspection, the architect will go through the job making notes (punchlist). In most cases he can't see how the wiring was installed. What he's checking are the things he can see. The lighting should be complete and installed according to the ceiling plan. The outlets should be located as shown on the drawings.

The quality and manufacturer of lighting fixtures and wiring devices are difficult to identify once the installation is completed. Most fixtures and devices have low-grade alternates that look just like the original. Grades and sizes of other materials are also hard to determine after they've been installed.

The number of lighting fixtures and wall outlets connected to a specific circuit is difficult to check. The electrical contractor may have changed the wiring scheme to eliminate certain circuits from the system. Those making the inspection rarely check this. If lighting fixtures and other equipment work properly, the job seems to be complete.

I'm not advocating that you cut corners on your jobs. But you should be aware that it is done and recognize that some contractors bid the job knowing that they won't complete the work as specified.

As-built Drawings

Fortunately, most contractors and subcontractors follow the plans carefully. If a problem arises during the installation, the architect will decide on the corrections to be made. If the contractor or a subcontractor deviate from the plan, the architect can order correction at no cost to the owner.

When changes are made at the job site, those making the change must coordinate it with others working on the job.

When these minor changes are made to the work, they should be noted on a set of drawings used for this purpose. When the job is completed, all other changes are noted on these drawings. They're marked "As-built" and returned to the architect.

If other alterations are planned at a later date, the as-built drawings become very important. The designers for the alteration plan the work from the as-built drawings. If these drawings aren't correct, serious problems can result.

The original contractor may be selected to do the alteration. If one of his subcontractors re-engineered the job to cut costs without indicating the changes on the as-builts, that subcontractor may not be awarded the work. If he is awarded the alteration contract, even he may have problems making the necessary adjustments. Few owners would agree to a change order to cover his own discrepancies.

Value Engineering

Value engineering is offered in some contracts as a way of encouraging the contractor to reduce costs. Most government contracts include a value engineering incentive provision.

Value engineering provisions usually apply when any contractor-developed proposal reduces the contract price. The following are typical requirements for most value engineering proposals:

• A description of the difference between the existing requirements specified in the contract documents and the proposed changes, the advantages and disadvantages of each, justification when an item's function or characteristics are being altered by the change, and the effect of the change on system performance.

• A list of the contract requirements that must be changed, and suggested specification revisions for the changed item.

• A detailed cost breakdown for both the original requirement and for the suggested change. The cost breakdown must include the contractor's implementation and development costs, including any amount that will go the subcontractors. The contractor must provide a breakdown of costs that the owner might incur in approving the value engineering proposals. This breakdown has to include all testing, evaluating, operating and support costs.

• A summary of any effect the value engineering proposal could have on the owner's collateral costs.

• An estimate of the time needed to implement the proposal. The statement should include any effects of changes on the contract completion date or the established work schedule.

• If known, a list of previously approved value engineering plans.

• Submittal by the contractor of the value engineering details to the owner's representative for approval. Usually the owner must approve the submittal within 30 days. If additional time is needed, the owner should notify the contractor within the 30-day period, provide the contractor with reasons for the delay, and give the expected date for the decision. The owner should review all value engineering submittals promptly. But the owner usually isn't liable for construction delays while acting on submittals.

• The contractor may withdraw any value engineering submittal, in whole or in part, after the stipulated number of days listed for approval.

• The owner should give the contractor written notice if the submittal is not approved. The notice should state the reasons for rejection.

• The owner will issue a modification to the contract for submittals that are approved. The modification will include the change in contract price and any necessary changes to the completion date.

• The owner may issue a notice to proceed with the value engineering proposal and negotiate a change in contract price and completion date.

• The owner's decision is final and not subject to the contract provision for disputes.

• The contractor may not proceed on any value engineering submittal until approved in writing by the owner.

• The contractor's portion of the value engineering savings may be determined by subtracting the owner's costs, then multiplying the balance by 60 percent.

Most contractors provide for value engineering cost reduction incentives in their subcontracts. Usually they negotiate settlements with subcontractors who submit value engineering cost reduction plans. The owner generally doesn't participate in these negotiations.

The contractor may restrict the owner's right to use any part of the value engineering plan or supporting data. This is usually done by placing a statement like the following in the submittal:

The breakdown data furnished under the Contract provisions for Value Engineering Cost Reduction Incentive shall not be disclosed outside of this Contract by the Owner, or duplicated, used or disclosed in whole or in part for any purpose other than to evaluate the submittal under this Contract provision. This restriction does not limit the Owner's right to use information from this submittal if it has been obtained or is otherwise available from the Contractor or from another source without limitations.

When the value engineering proposal is accepted, the contractor grants the owner unlimited rights for all data in the submittal, except technical data used to qualify or support the submittal. However, the owner has the right to specify and mark the data when making the approval.

When computing the contract savings, the contractor must not include any value engineering cost reduction payments made to subcontractors. These amounts are an implementation cost to the contractor.

Watch For Opportunities

Be alert to possible incentives. As an electrical estimator, you're in an excellent position to find ways of saving money by re-engineering. Even when the contract doesn't have a value engineering provision, you can still negotiate for one.

Most owners welcome cost reduction ideas and are willing to share the savings, which could be substantial. Cost reduction incentives also take the gamble out of quietly re-engineering the job to recoup a bid price that was a little too low.

Anyone putting up a building or remodeling one faces rising construction costs. A good cost reduction proposal should be well received, provided it doesn't reduce the quality or utility of the job.

Be sure the architect will like your proposal, too. The owner will most likely turn the proposal over to him for review and recommendations. Most architects don't like to change the appearance of jobs they've designed.

Value engineering provisions vary. Some have different computing schedules. Others accept only a minimum number of cost saving submittals to avoid a flood of minor proposals that could disrupt the construction schedule.

A value engineering proposal that reduces construction costs and results in a more useful structure has a good chance of being approved.

Chapter 16

Effects of Overtime

Overtime is time worked beyond the normal work day or week. There are many interpretations of what is considered "normal." Most employers observe an eight-hour day and a five-day week. Some use four ten-hour days; others follow a thirty-two- or thirty-five-hour work week.

Overtime pay is the rate paid for hours worked past normal working hours. Overtime pay is usually set at time-and-a-half or at double-time. Occasionally, more than the double-time rate is paid.

Some public works projects prohibit all overtime work and assess penalties if such work is done. The amount of the penalty is listed in the contract documents.

Always check for provisions that restrict job site operations. Sometimes overtime is necessary because of interruptions, for example. Some operations might require a shift longer than eight hours. Some installations must be done on weekends to avoid interfering with the daily activities of those occupying the area where the work is being done.

Efficiency and Fatigue
Establish the amount of overtime work required. Then adjust the labor rate on the take-off to include the overtime rate. And be sure to include a *fatigue factor*. Simply adding the increased labor rate to the normal manhours doesn't allow for fatigue. Your allowance for fatigue should be based on the reduced productivity you can expect from your crews.

Occasional overtime work can increase performance. There's more room to work and less interference when all other trades have left the job site at the end of the day. But studies by both trade associations and the federal government show that, generally, overtime reduces productivity. Accidents, absenteeism and mistakes increase, morale decreases, and employees become irritable and edgy. The problem of fatigue increases with the amount of overtime worked.

A widely used survey on the effects of overtime is Bulletin 917, prepared in 1947 by the Bureau of Labor Statistics. A chart from this survey, *Productivity as a Function of Work Days and Work Hours Per Day* (Figure 16-1) shows the effect on productivity of extended work days for a five-day week, a six-day week and a seven-day week. It also shows how productivity is affected when the work day is increased above the eight-hour standard. You may want to use figures from this chart when

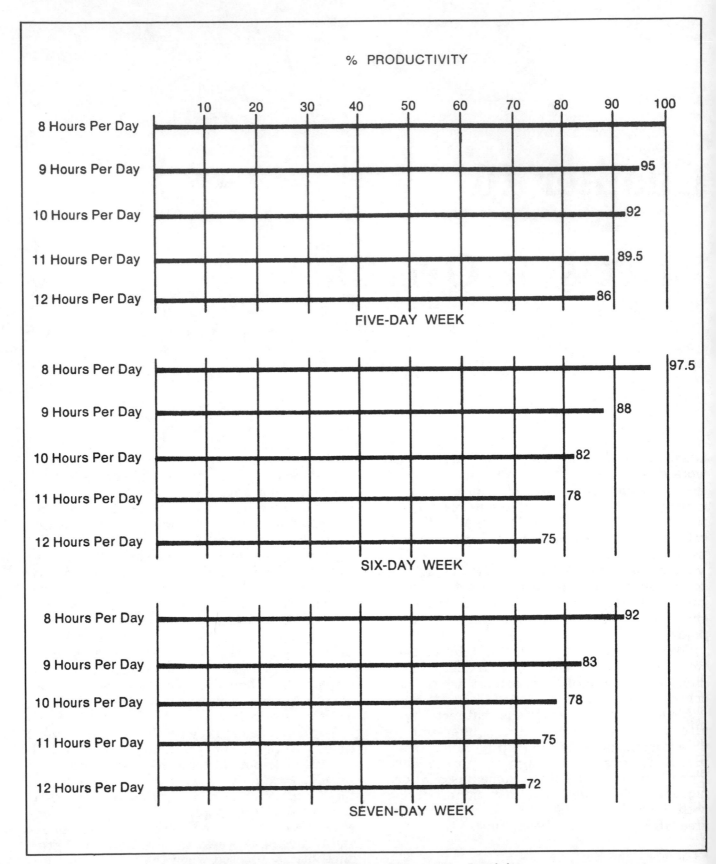

Effects of Extended Work Time on Productivity
Figure 16-1

calculating the lost productivity you can expect when overtime is needed.

The most extensive studies on the effects of overtime have been made by large manufacturers. These studies have found that, no matter what the employee incentive is, fatigue reduces productivity. And they've shown that the fatigue produced by overtime work affects work done on regular time, as well.

Trade associations have also made studies on overtime. Their findings are remarkably similar to those of Bulletin 917.

Most electrical jobs are not performed to the beat of a machine. The installation is usually a step-by-step process. But fatigue and inefficiency can still creep in. Your job as an estimator is to recognize that productivity is lower on overtime work and make an estimate of how that fact changes the job cost. Factor labor hours to allow reduced performance.

Most surveys indicate that working anyone seven days a week is foolish. The overall effect on performance of working a seventh day is usually counterproductive.

When the project schedule has been changed to rush completion and overtime is needed, you are entitled to more money for the remaining work. Any additional costs to you should be invoiced as extra work. It's as if you were issued a change order that added more work to the contract.

Most contractors and subcontractors find overtime costs excessive. It's to your advantage to avoid overtime when possible. When your crew falls behind, adding a few more employees may be the best answer. But if no qualified tradesmen are available, overtime may be the only answer.

Remember, though, that placing more employees on a project can also reduce productivity. Generally, the larger the crew size, the lower the production rate. This is where good supervision can really be an asset. You must be able to use the extra men effectively. Here are some points to consider before hiring extra men: What materials are available at the job site? Are there enough to get the job done? How about company tools? Are there enough to go around? Is there enough room for more employees on the job? Can you or your supervisor handle the additional manpower? Are the

other trades on the job going to increase work to meet your schedule?

Record Keeping

To negotiate a reasonable cost for the loss of productivity, you need recent, accurate and detailed records showing your company's experience. Your case is very weak without the figures and documents which support your claim.

Contractors and owners seldom recognize additional labor costs resulting from fatigue. You'll need to show them logical, well-written, accurate records. Try to establish a formula for covering the necessary factored labor units.

Develop a formula for overtime work by keeping detailed records of completed jobs and jobs in progress. Each company has its own system of record keeping. You'll have to determine what's best for you. The following are common records kept by many contractors:

• A daily, well-maintained job diary listing all pertinent data, both good and bad.

• Job codes for each major segment of work to be performed.

• Comparisons of estimated coded segments and the actual hours of labor used.

• Records of job overrun, both in days and manhours, beyond the original contract date.

• A close accounting on all overtime hours on each job.

After a while you'll see a pattern develop. The records will show whether your standard labor units are enough. They'll show the performance of each crew and supervisor and pinpoint areas that run up your costs. They'll indicate the type of work best suited to your company. And they help prove that overtime work has reduced productivity.

Your records will quickly show which contractors cost more to work for. If they have poorly-executed schedules and lack good, progressive supervision, you'll suffer the losses. A general contractor that delays the job and runs up your costs should pay more for your services.

If you don't keep good records, there's no way to convince contractors and owners to honor claims for additional costs. Even with good records, selling a claim is tough. No one wants to fund your misfortunes. But good records are hard to argue with.

Chapter 17

Record Keeping

The best cost data available to you will always be your own cost records. These records should include the following:

- The Estimate Log
- Labor Unit Bank
- Job Code Form
- Contractor Awards
- Job Log
- Daily Diary
- Job Analysis Record
- The Bid Schedule

The Estimate Log
Use a simple bound notebook with lined pages for your estimate log. This log becomes a history of the jobs you've bid.

Establish a logical and consistent method of identifying your bids. Numbering each estimate helps distinguish it from other office records. And be sure to record the name and basic data for each job selected for bid.

Prepare a file folder for each estimate. The label on the folder should include the estimate number, job name, date, and time the job is scheduled to be bid. Some estimators use stamps with this and additional information. A stamp like the one in Figure 17-1 will help you remember cost items you might overlook.

| Owner _____ |
| Contractor _____ |
| Architect _____ |
| Elec. Eng._____ |
| |
| Bid Bond   Yes   No |
| Performance Bond   Yes   No |
| Plan Deposit $_____ |

Figure 17-1

Listing the contractor on the file folder lets those concerned know immediately that the job is negotiated rather than bid. Identifying the architect is also important. During the prebid conference this information is available at a glance.

Labor Unit Bank
A labor unit bank will help you apply labor units to a job. The bank should list as many items of material as needed for the type of work being contracted. Each item will have a labor unit applied to it.

| | Per 100 Feet | | | Per Each | | | |
	Embedded	Exposed	Concealed	Connector	Elbow	Strap	Coupling
1/2"	3.25	3.50	3.25	.05		.05	.05
3/4"	3.50	4.00	3.75	.06		.06	.06
1"	4.00	4.50	4.25	.08		.08	.08
1-1/4"	4.50	5.00	4.75	.10	.10	.10	.10
1-1/2"	5.50	6.00	5.75	.10	.10	.10	.10
2"	7.00	8.00	7.50	.15	.15	.10	.15
2-1/2"	9.00	10.00	9.50	.20	.15	.10	.20
3"	10.00	12.00	11.00	.25	.20	.10	.25
4"	12.00	14.00	13.00	.25	.20	.10	.25

Electrical Metallic Tubing (EMT)

Labor Unit Bank
Figure 17-2

Use an inexpensive, bound notebook with lined pages. Divide each page into columns and list the materials and labor units, as shown in Figure 17-2. The labor unit should be listed next to the material item. Place a heading above the column to identify the material grouping.

Include pages for conduit, fittings, wire, boxes and wiring devices. Leave extra pages between the material groupings for additions.

Your labor units will never be perfect. They always need refining and additions. Set up your labor unit bank in such a way that changes can be made easily, without destroying the whole page.

Experienced electrical estimators are always trying to adjust labor units for new materials and procedures. New wiring methods require new labor unit categories. Be flexible with labor units; adjust them to fit the application.

Job Code Form

Develop a job code form that is simple, effective and easy to use. (See Figure 17-3.) It can be used as frequently as needed: by the day, week or month.

The job foreman lists the job code on weekly time cards for the work performed. The office posts the reported information to job accounting records. Reports on job progress are prepared from this data. Installation problems become obvious in a week or two rather than at the end of the job.

The job code form should not be burdensome and complicated. Field personnel shouldn't have to handle a lot of unnecessary and difficult paperwork. Their main job is getting the work finished. Keep the form simple.

After the contract is awarded, list the estimated manhours next to each coded item on the job code form. A secretary simply inserts the number of manhours used for the work period and calculates the balance that remains for each item. A copy should be sent to the job foreman as soon as it's prepared. It will help him detect problems while there is still work left to be done.

Job Code Form

Job Name **ADVERTISING COMPANY**   Job Number: **M-44**
Date: _____

Job Code	Estimated	Used	Balance
1. Switchgear	53		
2. Lighting fixtures, lamps	88		
3. Conduit, fittings	308		
4. Feeder conduits, fittings	12		
5. Wire	62		
6. Feeder cable	21		
7. Wiring devices	57		
8. Motor control equipment	11		
9. Hookup, control wiring	10		
10. Trenching, excavation	8		
11. Concrete encasement	3		
12. Manholes, handholes	3		
13. Demolition, disconnect			
14. Miscellaneous	29		
15. Uncoded			
16. Supervision			
17. Change orders			
18. Standby			
19.			
20.			

Figure 17-3

You should get a copy of the report. It helps you discover established standard labor units that need adjustment. Don't change your labor standards after each report, but the results can be very useful in spotting outdated figures.

A second job code form can be used for reporting actual job labor costs. The form can be designed like the labor manhours form. Attach this report to the manhour report. Management, the payroll department, and your office are the only places where this report should circulate.

The accounting report might show a crew of one foreman, three electricians, one beginning apprentice and one laborer. The composite hourly rate for the six workers will probably be less than the composite labor rate you used in the estimate. Other work periods will have different crew sizes and classifications, resulting in a different composite hourly rate.

Contractor Awards Record
Set up a file for contractor awards. This will help determine the awarding practices of the contractors you've bid. You may find that some contractors favor certain electrical contractors. If you're not one of their favored few, you may want to skip bids on their jobs in the future.

Arrange the log alphabetically, listing the contractor's name at the top of the page. Leave at least one full page for each contractor. List the jobs by date, and note the name, size, and type of job.

The awarding habits change. New personnel may go to work in the estimating department. Try to find out who they are. Be aware of their likes and dislikes. Some contractors won't tolerate an employee who plays favorites and upsets subcontractor bidding.

Job Log
A job log should record all change orders, job meetings, conversations and other useful information on a project. Include backup data for all changes. Indicate when and why job meetings were held. Make notes of materials, methods, or equipment that made the job easier or more difficult. And show how much pressure was necessary to get suppliers to fill purchase orders on time.

The log should show the steps taken to prepare change orders. This information is essential for relating details at the change order review meetings. A well-written backup and good record keeping will win many change order battles.

Daily Diary
It's good practice to keep a daily job diary. Most stationery stores sell a daily diary, but a simple bound notebook will do. Enter the activities of each work day in detail. Note important telephone conversations as well.

Be as detailed as you think necessary. And be sure to make your entries daily. It's hard to make an accurate entry for something that happened several days ago.

Daily diaries are valuable. Courts have accepted accurate, well-written diaries as evidence. Arbitrators also value them. A good diary can be the difference in winning or losing a dispute.

Job Analysis Record
Use a bound notebook for your job analysis record. List certain classes of work such as schools, churches, hospitals, industrial plants, warehouses, street lights and commercial housing units.

Write the type of construction at the top of the page. Then list the statistical data. (See Figure 17-4.) The data can be used to establish comparison costs and to determine preliminary project costs. The preliminary costs are an early check on the estimate.

Hospitals have many specialized requirements that need careful analysis. Hospitals and hospital additions have operating rooms, intensive care units, food handling areas, laundry and central building services. Some hospital additions have only patient rooms and none of the special purpose units. Be sure your analysis record includes a good description of any hospital job.

List job categories to make cost comparison easier. Your record should show the actual cost compared to the estimated cost. Provide another column for change orders, if needed.

The Bid Schedule
Like the other records, the bid schedule log can be an inexpensive notebook with lined pages. The bid schedule log has all your estimates, identified by the estimate number, job name and bid date. (See Figure 17-5.)

Estimate identification will vary from one company to another. Use whatever method is

Hospitals

Proj. Name Price	Contractor Name S.F.	Constr. Time Price	Total Bldg. S.F.	Total Bldg. Cost	Electric Contract S.F.	% of Job	Est. Per	Electrical Project	Electrical Cost
Rice	Simpson	18 Mos.	3,340,000	80,000	491,200	14.7	6.14	526,000	6.58
Mercy	Williams	24 Mos.	2,000,000	55,000	256,000	12.8	4.66	271,000	4.93
University	Cole	12 Mos.	550,000	10,000	61,000	11.2	6.16	68,800	6.88
County	Simpson	18 Mos.	1,500,000	30,000	187,500	12.5	6.25		

Job Analysis Record
Figure 17-4

M1	Johnson Motors	2/18/75
M2	Plaza Theatre	3/11/75
L1	Casey Electronics Plant	3/11/75
S1	Homer's Garage	5/14/75
L2	Sweet's Cannery	1/11/76
M3	Lovin' Kitchen	3/18/76
M4	Children's Hospital Addition A4	8/12/76
M5	Smitty's Appliance	2/14/77
S2	Central Meat Market	8/11/77
M6	Traffic Courthouse	5/16/78

Bid Schedule Log
Figure 17-5

best for you. The one shown below is simple and easy to use:

S—Small jobs
M—Medium jobs
L—Large jobs

Once you've chosen a method, stick with it. Mark all material pertaining to an estimate with the proper identification.

A *bid reporting system* can accompany the bid schedule. It contains followup data collected after the bid has been made. You will probably want to call contractors after the bid date and ask for a summary of the electrical bids. This tabulation is transferred to the contractor awards record. (See Chapter Twelve, Figure 12-1.) The rest of the data can be inserted in the bid schedule or the estimate log.

Chapter 18

Scheduling

Once you're awarded the job, it's time to begin planning. One of the first important steps is scheduling the flow of material, equipment and labor needed to perform the work. If you are like most estimators, you will be very involved with planning every step of the new project. Your first planning step will probably be ordering materials.

Material and Equipment Purchasing

You know what the material and equipment purchasing includes: Purchase orders must have enough information to ensure that the order is complete and meets contract requirements. They should include the project's address and point of delivery, the date that submittal data is required, the delivery dates and the terms of the sale.

Many electrical contractors have purchasing agents or buyers who do the actual purchasing from the supplier. You may only need to provide the purchasing agent with your material and equipment take-off. But the purchasing agent should have the name of the best supplier and his quote.

For example, a purchase order for lighting fixtures should give the fixture type, number of fixtures for each type, the voltage, and such items as plaster frames, end caps, stem and canopies, couplings, spacers, and special ballasts. It should mention that all material and equipment must comply with the specifications and drawings for the project. A typical purchase order is shown in Figure 18-1.

All material and equipment to be installed during the rough-in phase should be scheduled for delivery early in the job. Those items include conduit, fittings, boxes and box fittings, floor boxes, sleeves, underfloor systems, flush panel cans, flush terminal cabinets, and wire.

Items that will be installed after the rough-in must be scheduled accordingly. Such items include surface lighting fixtures, lamps, panel interiors and trim, switchboards, and wiring devices.

Some items require a long time for fabrication and shipment. These include switchboards, motor control centers, and substations. Order such items early, and follow up your orders so you get timely delivery.

You may need to include some special notes on the purchase order. These might be specific job requirements located in a drawing note, on the lighting fixture schedule, or in the project specifications. Otherwise, the supplier won't include those requirements.

Purchase Order

No. 0000

Smith Electric
15 Main Street
Dayton, OH 44863
(523) 841-1883

To: Westside Elec. Supply Co.   Ship: Advertising Company
312 Chicago Street   To: Industrial Park, Unit A
Super City, USA   Super City, USA
Attn: Mr. Al Gartland   c/o Associated Contr., Inc.
Date 3/30   Terms 30 Days FOB Jobsite   Tax   X Yes   No
Job Number M-44

Qty	Description	Price
1	Type A Lighting Fixture	54.00/E
46	Type A1	31.50/E
4	Type B	193.00/E
1	Type C	55.00/E
4	Type D	40.00/E
4	Type E	35.00/E
1	Type E1	40.00/E
2	Type F	45.00/E
2	Type G	75.00/E
2	Type H	40.00/E
2	Type J	25.00/E
10	Type K	1045.00/Lot
4	Type L	30.00/E
12	Type M	3000.00/Lot
1	Type N	40.00/E

See delivery and submittal schedule which shall be as listed. All Items
shall be in strict accordance with plans & specs. Plainly mark all fixture
cartons as to indicated type.

Typical Purchase Order
Figure 18-1

Prepare several copies of each purchase order. Send the first copy and one acknowledgement copy to the supplier. Insist that the supplier sign the acknowledgement copy and promptly return it. If the supplier finds something wrong with the purchase order, he should immediately return the whole order for correction. He may return the signed acknowledgement copy along with a note listing parts of the order which are unacceptable.

The supplier may qualify his quotation by stating that the prices are good until a certain date, such as the end of the first quarter. The sample purchase order in Figure 18-1 specifies that delivery must be made no later than March 15, which is before the end of the first quarter. If the supplier stated that a two-percent charge would be added after the end of the first quarter and the delivery date requested is after the first quarter, then the price listed on the purchase order should reflect that addition.

Coded Labor Breakdown

You should be able to prepare a coded labor breakdown with little additional effort. Others in the office can do it, but they might have trouble identifying uncommon or unusual items.

Establish a coded labor breakdown that can be used on all types of jobs. Keep it simple so that it is easily understood. The system should be used every time so it is always current for up-to-date job analysis.

Figure 18-2 shows a typical labor code breakdown. The alphabetized code can be changed to a numbering system or a combination of letters and numbers. Whichever is used, keep it simple. If the code is written on the weekly time cards, it must be easy to use.

The project supervisor must be kept aware of the job's progress as it compares with estimate projections. If he sees problems developing, he can take corrective action to avoid possible delays.

Feedback from workers and others on the job should always be directed to you. This way you know if your estimates are accurate or if adjustments are needed.

Project Construction Schedule

The general contractor is responsible for establishing and maintaining the total con-struction schedule. His goal is to ensure completion within the time limits set by the contract. The contractor will call on the major subcontractors to estimate lead time needed to get special equipment. Installation times are also established and placed in the schedule.

You and your supplier may have to work up a lead time schedule that will fit the master schedule. You must also coordinate the time needed to install the various phases of the work.

There are many types of construction schedules, from the simple bar chart (Figure 18-3) to the more complex critical path method. Begin your scheduling as soon as the contract has been signed. List the submittal time for catalog data, shop drawings, material lists and samples. There should be a step-by-step listing of construction phases. This list should include rough grading, excavations, foundations, rough wall construction, flatwork, roof structure, block work, roofing, electrical rough-in, plumbing and HVAC rough-in, lathing or dry wall, ceiling preparation, electrical equipment, plumbing equipment, HVAC equipment, wall covering, ceiling materials, painting, finish electrical, finish plumbing, finish HVAC, carpet or floor coverings, landscape and site work finish.

The construction schedule is usually submitted to the architect or to the owner's representative. If the schedule seems reasonable and identifies all construction phases, it will be marked "approved" and returned to the contractor. The contractor will then send a copy to each subcontractor so that they can schedule their work.

Some contractors hire a firm that specializes in preparing construction schedules. Most of these firms produce well-detailed schedules showing how each phase meshes with others.

Schedule charts can range from a small single-page bar chart to a full-size 30" x 52" sheet. Even larger, more complex charts are used on very large projects. The chart is divided into columns for the days or weeks planned to construct the project. The left side of the page lists the categories of work to be performed. A heavy line extends from the date an activity must begin to the day it must be completed. (See Figure 18-3.)

Labor Code Breakdown

Job Name _____ADVERTISING COMPANY_____

Job Number __M-44__

Date _____

Item	Estimate	Used	Balance
A. Trenching, Excavation	7.00		
B. Concrete, Sand, Encasement	3.00		
C. Feeder Conduits	49.00		
D. Branch Circuit Conduit Underslab	67.00		
E. Branch Circuit Conduit, Boxes All Other Areas	193.00		
F. Feeder Wire and Cable	26.00		
G. Branch Circuit Wire	60.00		
H. Communications Cable			
I. Lighting Fixtures, Lamps	88.00		
J. Panelboards, Switchboards	53.00		
K. Substations			
L. Motor Control Centers			
M. Motor Control Devices, EXO	8.00		
N. Control Wiring			
O. Wiring Devices	59.00		
P. Change Orders			
Q. Uncoded and Miscellaneous	40.00		
R. Mechanical Equip. Hookup	12.00		
S. Supervision			
	665.00		

The alphabetized code can be changed to a numbering system or a combination of letters and numbers. Whichever is used, be sure to keep it simple. If the employees are to indicate the code on weekly time cards then the simplier the code is, the easier to use.

Typical Labor Code Breakdown
Figure 18-2

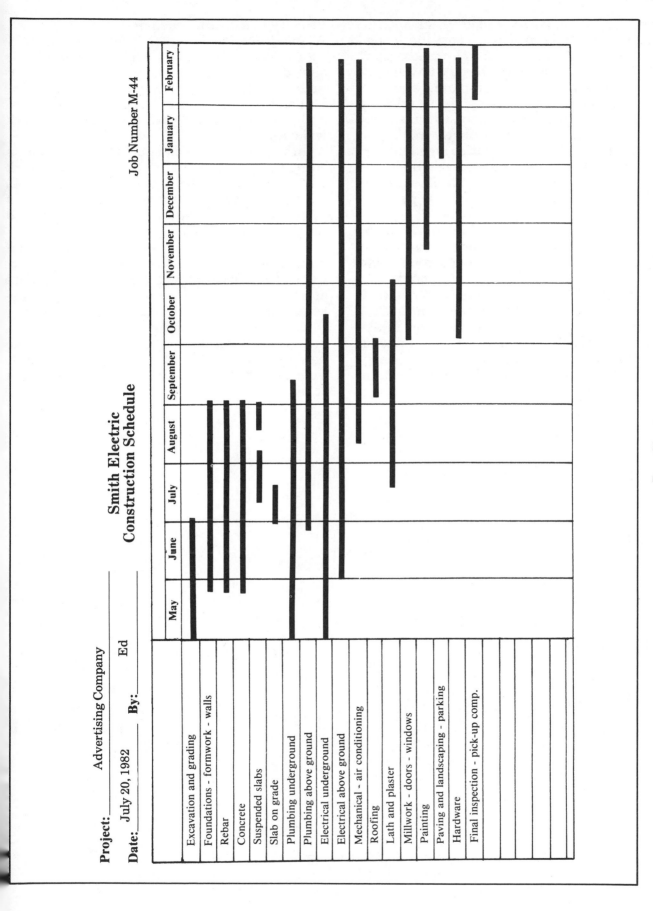

Bar Chart
Figure 18-3

Critical path charts are called PERT or CPM (Critical Path Method). These charts use coded activities placed over a daily or weekly chart that's similar to the bar chart. CPM begins with the time allowed for preparing and submitting material lists, catalog data and shop drawings for approval. It's good to have some background in CPM to properly read and understand the charts. Activities are identified on the chart by a circle with a code inside. A line is drawn to another circle. The line may be identified as a category of work to be performed at a specific time.

It's foolish to prepare a schedule without the advice of key subcontractors. Firms that specialize in making charts have a good working knowledge of how the chart should look, but few know the actual installation time needed. When a chart doesn't allow enough time for an installation, notify the contractor. Make sure that deliveries coincide with the installation date. To stay on schedule, the work may have to be started earlier than the chart indicates.

Progress Payment Schedule

A progress payment schedule is usually established immediately after the contract is signed. The owner needs to know how to arrange construction funding and the cash flow needed to service it. He'll also need to know exactly how much each major category of work will cost.

On a larger job, each subcontractor submits a breakdown of the major divisions of his work and the cost for each division. (See Figure 18-4. Sometimes the owner will demand that a prepared list of categories be priced and used for the progress billings.

The owner may want a more detailed breakdown for billing purposes. Or he may want the total electrical price as a lump sum amount. The contractor will inform each subcontractor of the content of the breakdown.

It makes bookkeeping much easier if the breakdown figures are rounded off to a whole number rather than given as the exact amount. This is standard practice on most payment schedules.

Progress Payment Form

ADVERTISING COMPANY

Date _____

Job Number __M-44_____

Request No. _____

Electrical Work   **Contract Amount:** $ __$40,965.00_____

1. Rough-in   15,965.00
2. Panels & Switchboard   7,000.00
3. Lighting Fixtures   12,000.00
4. Wiring Devices   2,000.00
5. Hookup Equipment   1,000.00
6. Sound System   3,000.00
7. Approved Change Orders
8. Total Amount
                         _____
                         40,965.00

The owner may want a more detailed breakdown that will be used for billing purposes or the Owner may simply just want the total electrical price as a lump sum amount. The Contractor will direct each Subcontractor as to the content of the breakdown.

It will make the following bookkeeping much easier if the breakdown figures are rounded off to a whole number rather than listing the exact amount. The rounded off amount is generally accepted by everyone involved.

Figure 18-4

Chapter 19

Information Sources

Every electrical estimator should be familiar with the resource materials and aids that are listed in this chapter. There's a wealth of information available to help you estimate job costs if you know where to look. And the more experienced you become as an electrical estimator, the more you'll realize that no one knows all there is to know about the subject. Every electrical estimator has to keep learning, finding new sources of information and benefitting from the experience of others.

You don't need to use every information source in this chapter. But at some time nearly every electrical estimator will rely on some reference listed here.

Trade Associations

Trade associations have been formed to set standards for estimating. They prepare bulletins and manuals on estimating techniques and have developed charts, diagrams, forms and other materials to help the practicing estimator make accurate, detailed estimates. Most offer membership to construction estimators. Some of the main ones are:

Association of Energy Engineers
4025 Pleasantdale Road, Suite 340
Atlanta, Georgia 30340

American Institute of Plant Engineers
3975 Erie Avenue
Cincinnati, Ohio 45208

American Society of Professional Estimators
5201 North 7th Street, Suite 200
Phoenix, Arizona 85014

Edison Electric Institute
111 19th Street N.W.
Washington, D.C. 20036

International Association of Electrical Inspectors
802 Busse Highway
Park Ridge, Illinois 60068

Institute of Electrical and Electronics Engineers, Inc.
345 East 47th Street
New York, New York 10017

Illuminating Engineering Society of North America
345 East 47th Street
New York, New York 10017

International League of Electrical Associations
2101 L Street, N.W.
Washington, D.C. 20037

National Association of Electrical Distributors

P.O. Box 5037
Stamford, Connecticut 06904

National Electrical Manufacturers Association
2101 L Street, N.W.
Washington, D.C. 20037

National Fire Protection Association
Batterymarch Park
Quincy, Massachusetts 02269

National Lighting Bureau
2101 L Street, N.W.
Washington, D.C. 20037

Trade Magazines

Many trade magazines feature articles on estimating. Some show cost breakdowns for completed projects along with pictures, illustrations and diagrams explaining the estimating procedure in step-by-step detail. There have been articles giving a detailed comparison of the estimated costs and final costs including change orders. Some even project the probable cost should the project be bid at the time of completion.

Trade magazines feature new products and applications, and carry advertizing of major products, tools, equipment, etc. You may contact the major publications by writing to:

CEE, Sutton Publishing Co., Inc.
707 Westchester Avenue
White Plains, New York 10604

Electrical Business
Circulation Department
2506 Gross Point Road
Evanston, Illinois 60201

Electrical Contractor
7315 Wisconsin Avenue
Bethesda, Maryland 20814

Electrical Construction and Maintenance
P.O. Box 510
Hightstown, New Jersey 08520

Engineering News-Record
1221 Avenue of the Americas
New York, New York 10020

Trade Shows

You can find valuable information at trade shows sponsored by suppliers' associations. Many new and useful products are displayed, and salespeople or factory representatives are on hand to discuss their product lines. Information about local or regional trade shows is available from your supplier.

Some suppliers set up displays and provide literature on new products at their offices. There's usually a factory representative and a salesperson to offer assistance and discuss installation techniques. They can order special catalogs or brochures from their main office, if necessary.

Catalogs Catalogs are probably the best immediate source of information for the estimator. Keep catalogs up-to-date with the latest inserts and replacement sheets. They're usually free to the electrical contractor. Suppliers and manufacturer's representatives may visit your office to up-date their catalogs.

A well stocked catalog library will include the following:

Acme Lighting
ALR
Appleton
Arrow-Hart
Associated Products
Automatic Switch
Atmos
Belden
Bell
Bieber
Bowers
Bryant
Burndy
Buss
Cadweld
Caine-Strut
Capri Lighting
Carlon
Carol
Chance
Circle AW
Circle Wire
Cole
Columbia
Continental Wire
Cresent Lighting
Crouse-Hinds
Cutler-Hammer
Daybrite
Devine
Dynaray
Efcor
Elastimold
Elcen
Faraday

Federal Pacific Electric
Federal Signals
Gamewell
General Electric Supply
Globe Lighting
Graybar
Greenlee
Guth Lighting
G&W Products
Halo Lighting
Holophane
Hubbell
Hunt
Hydrel
ITT
Jet-Phillips
Jones-Laughlin
Kaiser
Keystone Lighting
Keene Lighting
Killark
Lawless Detroit Diesel
Kosman Lighting
Line Material
Leviton
Lithonia Lighting
Lumi-Dyne
Lutron
Marco
Matra
McGee
McGraw Hill
Minerallac
3-M
Miller
Minarik Controls
Moldcast
Mono
O-Z Products
Paragon
PLM Products
Prescolite Lighting
P&S Devices
P&W Products
Pyle-National
Quality Lighting
Raco
Reloc
Rob-Roy
Rome Cable
Russell-Stoll
Sangamo
Sierra
Shawmut Fuses
Slater

Solar Lighting
Siltron
Spaulding
Standby Systems
Supreme Lighting
Super-Strut
Sylvania
Thomas & Betts
Thermador
Tork
Triangle
Unistrut
Walker/Parkersberg
Western Plastics
Westinghouse
Wiremold
Widelight
Wiegmann
Woodhead
Yorklite
Zenith

Have your contacts at the suppliers or distributors help you in obtaining catalogs for your library. You can check the telephone directory for manufacturer's representatives and request catalogs. Visit the trade shows and be sure to sign up for catalogs at the manufacturer's booth or table top display.

New Products
When a new product or procedure shows up in a set of plans and you're unsure of the proper application, call the supplier. He'll probably have the information you need. If not, he can call the factory rep. Watch the new product being installed, and get some feedback from your supervisors and installers.

For each project, study the material and equipment required to determine the necessary labor values for making the installation. You may be a little high or a little low at first. But as you gain experience with the new item, your accuracy will increase.

Visit your local supplier occasionally. He probably has pamphlets on special products that he'll gladly give you. Meet his sales staff, and don't hesitate to call them for information or assistance.

Many suppliers use charts to compare the products of various manufacturers. Figure 19- is a chart comparing panelboards.

Trade Meetings
In some areas local trade associations have periodic meetings, with many industr

Panelboard Comparison Chart

Gen. Electric	Westinghouse	ITE	Square D	Kinney	Cutler-Hammer	Federal	Frank Adams
NLAB	NQB	NLAB	NQOB	NQB	NLAB	NBLP	QS
NLAB	NQB	NLAB	NQOB	NDQB		NBLP	
NLTQ	NQP	NPAB	NQO	NQP	NPLAB	NALP	QP
DB	NQP UNASSEMBLED	NPA	NQO UNASSEMBLED	NSQP	CHP	NALP	QPFPB
NAB	NEB	NAB	NAIB	NAB	NAIB	NAIB	NAIB
NHB	NHIB	NHB	NYIB	NHB	NHIB	NHIB	NHIR
CCB	CDP	CDP	ML	CDP	CDP	CDP	CDP
NAB	NEB	NDP	NAIB	NDP250	NDP	NHDP	NDP
NHB	NHDP	NHDP	NYIB	NDP480	NHDP	NHDP	NHDP
NHB				NDP600			
				SF250		FLEXUNIT	PFS, KSF
				SF600		FLEXUNIT	
				NSF250		FLEXUNIT	PFX, KSF
QMR	FDP	VB30,32	QMB	QMQB250			S-A-W
QMR	FDP	VB30,32	QMB	QMQB600		QMQB	S-A-W
		VB23		NOMQB250		NARROW QMQB	S-A-W
				NOMQB600			S-A-W
	NTIP,C	NTP,C		NTP,C		NTIP	LNTP
				T2P,C			
				NIP,C			NIP
CCB WITH PANEL MOTOR CONTROL UNIT	MSAB	MSC	MOTOR STARTER PANEL	CDP-MS	QVB	CIRCUIT BREAKER WITH PMS	MOTOR STARTER PANEL
QMR WITH PANEL MOTOR CONTROL UNIT	MSFS	MSF	MOTOR STARTER PANEL	QMQB-MS	QVB	FUSIBLE WITH PMS	MOTOR STARTER PANEL
NLTQ	NQP	NPAB	MQO			NALP	
			HCN, HCM				
			NTFB, NTHB				

Panelboard Comparison Chart
Figure 19-1

representatives attending. The aim of these gatherings is to gain better understanding and cooperation among the various trades. Local inspection agencies can answer questions regarding acceptable installation methods and compliance standards. Factory reps may have questions for the inspectors or vice versa. Usually, everyone in the industry is welcome.

Watch for any other gatherings, seminars or conventions that may involve the electrical industry. The local utility company may offer meetings or seminars to upgrade industry standards. Other trade societies, such as the Specification Writers (CSI) or the Illuminating Engineers Society, may have special meetings, inviting industry members to attend.

Education

Education is an important ingredient for success in any field, and electrical contracting and estimating is no exception. As the industry changes, so must you. If you don't, you'll be left behind. There's always something new to learn; every new job has some new twist or challenge. Most are minor and have little effect by themselves, but collectively they can make quite a difference.

Most estimators have a high school education and some years of experience in electrical installation. Additional education on subjects related to electrical estimating is very useful. Check with local high schools, colleges and trade associations for classes dealing with such topics as the building code, job site supervision, project scheduling and business management.

In some areas there are special courses on electrical estimating. Some colleges and universities offer seminars and extension courses on the subject.

Stay current on code changes that will affect you. Some code changes affect the final cost of a job. The code is often amended to include new products. If you're not aware of code changes concerning a new product or installation method, you could be missing out on the benefits these new items have to offer. Also, you could be using them in violation of code regulations.

Study and Discipline

There's no big secret to becoming a qualified electrical estimator. All it takes is care, skill, a willingness to learn and the desire to do the best job possible. But it does require discipline, hard work and study.

If you don't like reading or have difficulty following written instructions, estimating isn't for you. Reading specifications can be very boring, but it must be done. Sometimes you must dig deep into the documents to find the information you need. Every architect and electrical engineer has his own way of presenting a job. Some are clearer than others. Your take-off must be accurate, regardless of how the job is presented.

You must be able to perform the math required to compute costs. Machines can do the mechanical calculations, but it's up to you to provide the right data.

Learning to Use Computers

Computer estimating has been a big help to many estimators. But it hasn't replaced the need for good learning and reading skills. Computers only offer a quicker method for assembling the estimate package. They're very fast and mathematically correct. But for a computer to be useful in estimating, it must have an estimating program that fits your requirements.

A computer expert who doesn't know anything about electrical contracting can't write a good estimating program by himself. He'll need the help of an experienced estimator who knows something about computers and is able to test and modify the programs as needed.

Many community colleges offer courses on programming. A course in computer software can provide a basic knowledge of programming fundamentals. It may take a while to determine what your computer needs are and what the programs should do.

Once you've developed your own programs, it'll take time to get them up and running smoothly and effectively. Ask other electrical contractors and estimators in your area about computer estimating. They may have some good ideas.

Some computer companies offer seminars around the country. Usually these seminars are free to prospective customers. Of course, they're slanted towards the company's products.

Learning to be an Estimator

The education of an electrical estimator requires more than just an understanding of electrical work. You must be able to understand any part of the specifications and have a working knowledge of each trade involved. To determine how an installation must be made, you must understand the overall design of the area where the work will be done.

A basic education and the desire to continue learning are essential to becoming a competent and qualified electrical construction estimator.

Certified Estimator

The American Society of Professional Estimators (A.S.P.E.) is an organization of construction estimators that provides certification for qualifying candidates. It was established in the mid-1950's and began its certification program in 1976. Today, the A.S.P.E. is nationwide and has many members in countries around the world. The main office is at 5201 N. 7th Street, Suite 200, Phoenix, Arizona 85014, (602) 274-4880.

Members come from all areas of the construction industry. To apply for certification a member must have five years' experience in a specific trade. After your application is accepted, an examination is scheduled. The applicant must perform a take-off of a job in the trade he's applying for.

The examination is supervised by one of the organization's certified members. The prices and labor units supplied are used to arrive at a job cost. Certain lump sum prices may be used for special pricing that would normally be provided by a subcontractor or a supplier. The idea is that when the estimator uses the supplied plans and specifications, the projected electrical cost should fall within a certain range.

Examinations are scheduled periodically. If the applicant fails the first time, he may reapply. If he fails a second time, the applicant must wait for a certain period before another ex-

amination will be permitted.

The take-off plan and unit costs may vary from examination to examination, so you may not have the same one for the second try. The plan, specifications, labor units and pricing were all developed by the association. There's no way to brush up on the exam other than to be actively involved in estimating. The certification would be worthless if anybody could merely pick up a textbook and pass the examination.

The A.S.P.E. has a national office and local chapters in many major cities. Membership is limited to construction estimators or those closely allied with the industry. There are categories of membership, from student, to estimator-in-training, to estimator. The classification of "estimator" is for those having a minimum of five years' experience of active estimating.

The certification program was evaluated by two major universities at the time it was introduced. They determined that the testing was one of the most outstanding ever devised for the industry. Members of the organization who developed the certification program have many years of estimating experience. Some are educators who have taught construction estimating at the college level.

The tests are developed by estimators from many trades across the country and have been evaluated by the Society's members.

The certification is recognition that an estimator has been examined and found to be qualified. It informs others in the industry that the holder is a Certified Construction Estimator.

There are many benefits to being a member. You can learn a lot from associating with others in the same industry. Someone else has almost certainly handled nearly every problem you will face. Also, it's good to know as much about your competitor as possible. The A.S.P.E. promotes estimating as a profession and provides programs and seminars relating to construction estimating.

Chapter 20

Estimating Forms

The forms in this section are for your use. Copy them if you wish. Have an "instant printer" run off several hundred copies of any of these forms and make them into a pad for easy use. Some will have to be changed to fit your needs. I hope you'll find them useful in preparing a take-off or bid.

There are thirteen forms in this section:

1. Project Selection Checklist
2. Scope of Work
3. Work Sheet
4. Pricing Sheet
5. Bid Summary
6. Telephoned Quotations
7. Spread Sheet
8. Contractor Award Record
9. Bidding Scope Form
10. Job Code Form
11. Labor Code Breakdown
12. Progress Payment Form
13. Purchase Order

Project Selection Checklist

Job _____   Bid Date _____
Location _____   Estimator _____

Financial

1. Approximate Cost of Work _____
2. Bonding Required _____
3. Progress Payments _____
4. Retention _____%
5. Delay Penalties _____
6. _____
7. _____

Bid Documents

1. Complete Plans _____
2. Complete Specs. _____
3. Reduced Plans _____
4. Plan Deposit _____
5. Public Bid _____
6. Sublisting _____
7. Prequalify _____

Project Type

1. Residential _____
2. Commercial _____
3. Industrial _____
4. Institutional _____
5. Underground _____
6. Overhead _____
7. Waterfront _____
8. High Voltage _____
9. Communications _____
10. _____
11. _____
12. _____

Basic Factors

1. Firm Price _____
2. Negotiated _____
3. Special Equipment _____
4. Construction Time _____
5. Adequate Labor _____
6. Adequate Equipment _____
7. Adequate Tools _____
8. Site Conditions _____
9. Unusual Problems _____
10. _____
11. _____
12. _____

Scope of Work

Estimate No. _____

Job _____   Bid Date _____

Location _____   Estimator _____

Bids To_____ Location_____   Time _____

1. Design Team
Architect   _____
Engineer   _____
Agency   _____
Owner   _____

2. Construction
Building   _____
Walls   _____
Ceilings   _____
Floors   _____

3. Quotations
Switchgear   _____
Generator   _____
Alarm Systems   _____
U.F. Duct   _____
Communication Systems   _____
Cable Tray   _____
Fixtures   _____
Telemetry   _____

_____   _____
_____   _____
_____   _____
_____   _____

4. Specified Items
Conduit   _____
Wire   _____
Switches   _____
Receptacles   _____
Dimming Equipment   _____
Motor Control   _____
Manholes   _____
Concrete   _____

_____   _____
_____   _____
_____   _____
_____   _____

5. Related Work
Temporary   _____
Control Wiring   _____
Starters   _____
Painting   _____
Service Cable   _____
Pole Bases   _____

_____   _____
_____   _____
_____   _____
_____   _____

6. Site Conditions
Excavation   _____
Access   _____
Utilities   _____
Security   _____
Pave Cutting   _____
Pave Patch   _____
Demolition   _____

_____   _____
_____   _____
_____   _____
_____   _____

Work Sheet

Estimate No.:_____

Pricing Sheet

Job: _____   Estimate No.: _____

Work:_____   Sheet:_____of_____

Estimator: _____Checker _____   Date: _____

Description	Qty.	Price	Per	Extension	Hours	Per	Extension
				Total			

Bid Summary

Job: _____   Estimate No.: _____
Location: _____   Bid Date: _____
Division: _____   Time: _____
Estimator: _____   Checker: _____

	Description	Material		Deducts		Labor

	Material				Labor
Sub Total					
Sales Tax ____%				Supervision %	
Sub Total				Total Hours	
Labor				Rate	
Tools				Sub Total	
Miscellaneous				Fringes	
Permits/Fees				Taxes	
Sub Total				Sub Total	
Subcontracts				Factor	
Travel Expense				Total	
Sub Total					
Deducts				Addendums	_____
Net Cost				Alternates	_____
Overhead				Job Duration	_____
Sub Total				Penalty	_____
Profit				Type Const.	_____
Contingency					
Bond					
Selling Price					

Telephoned Quotations

Job: _____   Estimate No.: _____

Supplier: _____   Estimator: _____

Person Quoting: _____   Time: _____

Date: _____

Quantity	Description	Price	Net
		Total	

F.O.B. Jobsite _____   Includes:

Tax Included _____   _____

Installed _____

Plans & Specs. _____

Addendums _____   Excludes:

Division _____   _____

_____

_____   _____

_____

Spread Sheet

Job: _____   Estimate No.: _____

Material: _____   Date: _____

Contractor Award Record

Project	Date	Our Price	Awarded To	Price

Bidding Scope Form

Project Name: _____   **Bid Date:** _____

Location: _____

Section Bidding: _____   **Estimator:** _____

Scope:

Includes:

Excludes:

Job Code Form

Job Name _____ **Job Number:** _____

Date: _____

Job Code	Estimated	Used	Balance
1. Switchgear	_____	_____	_____
2. Lighting fixtures, lamps	_____	_____	_____
3. Conduit, fittings	_____	_____	_____
4. Feeder conduits, fittings	_____	_____	_____
5. Wire	_____	_____	_____
6. Feeder cable	_____	_____	_____
7. Wiring devices	_____	_____	_____
8. Motor control equipment	_____	_____	_____
9. Hookup, control wiring	_____	_____	_____
10. Trenching, excavation	_____	_____	_____
11. Concrete encasement	_____	_____	_____
12. Manholes, handholes	_____	_____	_____
13. Demolition, disconnect	_____	_____	_____
14. Miscellaneous	_____	_____	_____
15. Uncoded	_____	_____	_____
16. Supervision	_____	_____	_____
17. Change orders	_____	_____	_____
18. Standby	_____	_____	_____
19.	_____	_____	_____
20.	_____	_____	_____
	_____	_____	_____

Labor Code Breakdown

Job Name _____

Job Number _____
Date _____

Item	Estimate	Used	Balance

Progress Payment Form

_____ Date _____

Job Number _____ **Request No.** _____

Electrical Work   **Contract Amount:** $ _____

Purchase Order

No.

To: _____   Ship: _____
_____   To: _____
_____   _____
Attn: _____   _____
Date_____Terms_____FOB_____Tax_____Yes_____No
Job Number_____

_____
_____
_____
_____
_____
_____
_____
_____
_____
_____
_____
_____
_____
_____
_____
_____
_____
_____
_____
_____
_____
_____
_____
_____

Index

Other Practical References

Manual of Electrical Contracting

A detailed guide to the design and installation of electrical systems in residential, commercial and agricultural buildings. For both new construction and remodeling work, this practical manual provides you with the construction and business essentials required to make sure, before you even start your business, that it will succeed. Construction essentials include: drawing up electrical plans; designing the correct lighting; calculating service and feeder loads; service entrance capacity and demand factors, and system installation. Business essentials include: developing the necessary finances; staying within your budget; using a simple, well-organized system of record-keeping (sample forms included in text), and keeping your business running profitably. **224 pages, 8½ x 11, $17.00**

Electrical Blueprint Reading

Shows how to read and interpret electrical drawings, wiring diagrams and specifications for construction of electrical systems in buildings. Shows how a typical lighting plan and power layout would appear on the plans and explains what the contractor would do to execute this plan. Describes how to use a panelboard or heating schedule and includes typical electrical specifications. **128 pages, 8½ x 11, $8.50**

Electrical Construction Estimator

If you estimate electrical jobs, this is your guide to current material costs, reliable manhour estimates per unit, and the total installed cost for all common electrical work: conduit, wire, boxes, fixtures, switches, outlets, load centers, panelboards, raceway, duct, signal systems, and more. Explains what every estimator should know before estimating each part of an electrical system. **400 pages, 8½ x 11, $25.00**

National Construction Estimator

Current building costs in dollars and cents for residential, commercial and industrial construction. Prices for every commonly used building material, and the proper labor cost associated with installation of the material. Everything figured out to give you the "in place" cost in seconds. Many time-saving rules of thumb, waste and coverage factors and estimating tables are included. **512 pages, 8½ x 11, $16.00. Revised annually.**

Repair And Remodeling Cost Estimator

The complete pricing guide for dwelling reconstruction costs. Reliable, specific data you can apply on every remodeling job. Up-to-date material costs and labor figures based on thousands of repair and remodeling jobs across the country. Professional estimating techniques to help determine the material needed, the quantity to order, the labor required, the correct crew size and the actual labor cost for your area. **256 pages, 8½ x 11, $16.75. Revised annually**

Building Cost Manual

Square foot costs for residential, commercial, industrial, and farm buildings. In a few minutes you work up a reliable budget estimate based on the actual materials and design features, area, shape, wall height, number of floors and support requirements. Most important, you include all the important variables that can make any building unique from a cost standpoint. **240 pages, 8½ x 11, $12.00. Revised annually**

Construction Industry Production Manual

Manhour tables developed by professional estimators from hundreds of jobs and all types of construction. Thousands of carefully researched figures, accurate charts and precise tables to give the estimator the information needed to compile an estimate for residential and light commercial construction. **176 pages, 5½ x 8½, $8.00**

Contractor's Year-Round Tax Guide

How to set up and run your construction business to minimize taxes: corporate tax strategy and how to use it to your advantage, why you should consider incorporating to save tax dollars, and what you should be aware of in contracts with others. (Includes sample contracts). Covers tax shelters for builders, write-offs and investments that will reduce your taxes, accounting methods that are best for contractors, what forms of compensation are deductible, and what the I.R.S. allows and what it often questions. Explains how to keep records and protect your company from tax traps that many contractors fall into. **192 pages, 8½ x 11, $16.50**

Construction Estimating Reference Data

Collected in this single volume are the building estimator's 300 most useful estimating reference tables. Labor requirements for nearly every type of construction are included: site work, concrete work, masonry, steel, carpentry, thermal & moisture protection, doors and windows, finishes, mechanical and electrical. Each section explains in detail the work being estimated and gives the appropriate crew size and equipment needed. Many pages of illustrations, estimating pointers and explanations of the work being estimated are also included. This is an essential reference for every professional construction estimator. **368 pages, 11 x 8½, $18.00**

Residential Electrical Design

Explains what every builder needs to know about designing electrical systems for residential construction. Shows how to draw up an electrical plan from the blueprints, including the service entrance, grounding, lighting requirements for kitchen, bedroom and bath and how to lay them out. Explains how to plan electrical heating systems and what equipment you'll need, how to plan outdoor lighting, and much more. If you are a builder who ever has to plan an electrical system, you should have this book. **194 pages, 8½ x 11, $11.50**

Contractor's Guide To The Building Code

Explains in plain English exactly what the Uniform Building Code requires and shows how to design and construct residential and light commercial buildings that will pass inspection the first time. Suggests how to work with the inspector to minimize construction costs, what common building short cuts are likely to be cited, and where exceptions are granted. If you've ever had a problem with the code or tried to make sense of the Uniform Code Book, you'll appreciate this essential reference. **312 pages, 5½ x 8½, $16.25**

Rough Carpentry

All rough carpentry is covered in detail: sills, girders, columns, joists, sheathing, ceiling, roof and wall framing, roof trusses, dormers, bay windows, furring and grounds, stairs and insulation. Many of the 24 chapters explain practical code approved methods for saving lumber and time without sacrificing quality. Chapters on columns, headers, rafters, joists and girders show how to use simple engineering principles to select the right lumber dimension for whatever species and grade you are using. **288 pages, 8½ x 11, $14.50**

Wood Frame House Construction

From the layout of the outer walls, excavation and formwork, to finish carpentry, and painting, every step of construction is covered in detail with clear illustrations and explanations. Everything the builder needs to know about framing, roofing, siding, insulation and vapor barrier, interior finishing, floor coverings, and stairs. . .complete step by step "how to" information on what goes into building a frame house. **240 pages, 8½ x 11, $11.25 Revised edition**

Builder's Office Manual

This manual will show every builder with from 3 to 25 employees the best ways to: organize the office space needed, establish an accurate record-keeping system, create procedures and forms that streamline work, control costs, hire and retain a productive staff, minimize overhead, shop for computer systems, and much more. Explains how to create routine ways of doing all the things that must be done in every construction office in a minimum of time, at lowest cost and with the least supervision possible. **208 pages, 8½ x 11, $13.25**

Estimating Plumbing Costs

Offers a basic procedure for estimating materials, labor, and direct and indirect costs for residential and commercial plumbing jobs. Explains how to interpret and understand plot plans, design drainage, waste, and vent systems, meet code requirements, and make an accurate take-off for materials and labor. Includes sample cost sheets, man-hour production tables, complete illustrations, and all the practical information you need to accurately estimate plumbing costs. **224 pages, 8½ x 11, $17.25**

Builder's Guide to Accounting

Explains how to set up and operate the record systems best for your business: simplified payroll and tax record keeping plus quick ways to make forecasts, spot trends, prepare estimates, record sales, receivables, checks and costs, and control losses. Loaded with charts, diagrams, blank forms and examples to help you create the strong financial base your business needs. **304 pages, 8½ x 11, $12.50**

Construction Superintending

Explains what the "super" should do during every job phase from taking bids to project completion on both heavy and light construction: excavation, foundations, pilings, steelwork, concrete and masonry, carpentry, plumbing, and electrical. Explains scheduling, preparing estimates, record keeping, dealing with subcontractors, and change orders. Includes the charts, forms, and established guidelines every superintendent needs. **240 pages, 8½ x 11, $22.00**

Plumbers Handbook Revised

This 1985 edition shows what will and what will not pass inspection in drainage, vent, and waste piping, septic tanks, water supply, fire protection, and gas piping systems. All tables, standards, and specifications are completely up-to-date with recent changes in the plumbing code. Covers common layouts for residential work, how to size piping, selecting and hanging fixtures, practical recommendations and trade tips. This book is the approved reference for the plumbing contractors exam in many states. **240 pages, 8½ x 11, $16.75**

Basic Plumbing With Illustrations

The journeyman's and apprentice's guide to installing plumbing, piping and fixtures in residential and light commercial buildings: how to select the right materials, lay out the job and do professional quality plumbing work. Explains the use of essential tools and materials, how to make repairs, maintain plumbing systems, install fixtures and add to existing systems. **320 pages, 8½ x 11, $17.50**

Masonry & Concrete Construction

Every aspect of masonry construction is covered, from laying out the building with a transit to constructing chimneys and fireplaces. Explains footing construction, building foundations, laying out a block wall, reinforcing masonry, pouring slabs and sidewalks, coloring concrete, selecting and maintaining forms, using the Jahn Forming System and steel ply forms, and much more. Everything is clearly explained with dozens of photos, illustrations, charts and tables. **224 pages, 8½ x 11, $13.50**

Handbook of Modern Electrical Wiring

The journeyman electrician's guide to planning the job and doing professional quality work on any residential or light commercial project. Explains how to use the code, how to calculate loads and size conductors and conduit, the right way to lay out the job, how to wire branch and feeder circuits, selecting the right service equipment, and much more. **204 pages, 5½ x 8½, $14.75**

Craftsman BOOK COMPANY

6058 Corte del Cedro
P. O. Box 6500
Carlsbad, CA 92008

Phone Orders

For charge card orders call (619) 438-7828
Your order will be shipped within 48 hours of your call.

Name _____

Company _____

Address _____

City _____ State _____ Zip _____

Send check or money order
Total Enclosed _____ (In California add 6% tax)
If you prefer, use your ☐ Visa or ☐ MasterCard

Card no. _____

Expiration date_____ Initials_____

Mail Orders
We pay shipping when your check covers your order in full.

10 Day Money Back GUARANTEE

- ☐ 17.50 Basic Plumbing with Illustrations
- ☐ 12.50 Builders Guide to Accounting
- ☐ 13.25 Builders Office Manual
- ☐ 12.00 Building Cost Manual
- ☐ 18.00 Const. Estimating Ref. Data
- ☐ 8.00 Construction Industry Production Manual
- ☐ 22.00 Construction Superintending
- ☐ 16.25 Contractor's Guide To The Building Code
- ☐ 16.50 Contractor's Year-Round Tax Guide
- ☐ 8.50 Electrical Blueprint Reading
- ☐ 25.00 Electrical Construction Estimator
- ☐ 17.25 Estimating Plumbing Costs
- ☐ 14.75 Handbook of Modern Electrical Wiring
- ☐ 17.00 Manual of Electrical Contracting
- ☐ 13.50 Masonry & Concrete Construction
- ☐ 16.00 National Construction Estimator
- ☐ 16.75 Plumbers Handbook Revised
- ☐ 16.75 Repair And Remodeling Cost Estimator
- ☐ 11.50 Residential Electrical Design
- ☐ 14.50 Rough Carpentry
- ☐ 11.25 Wood-Frame House Construction
- ☐ 19.00 Estimating Electrical Construction

These books are tax deductible when used to improve or maintain your professional skill.

6785